Progress in Optical Science and Photonics

Volume 33

Indexed by Scopus

The purpose of the series Progress in Optical Science and Photonics is to provide a forum to disseminate the latest research findings in various areas of Optics and its applications. The intended audience are physicists, electrical and electronic engineers, applied mathematicians, biomedical engineers, and advanced graduate students.

Rajesh Sharma

Nonlinear Optical Materials for All-Optical Switching Applications

Springer

Rajesh Sharma
Department of Pure and Applied Physics
Guru Ghasidas Vishwavidyalaya
Bilaspur, Chhattisgarh, India

ISSN 2363-5096 ISSN 2363-510X (electronic)
Progress in Optical Science and Photonics
ISBN 978-981-96-5920-3 ISBN 978-981-96-5921-0 (eBook)
https://doi.org/10.1007/978-981-96-5921-0

This Springer imprint is published by the registered company Springer Nature Singapore Pte Ltd.
The registered company address is: 152 Beach Road, #21-01/04 Gateway East, Singapore 189721, Singapore

Dedicated to my mother and GOD, whose love and unwavering support have been my greatest strength.

And to all those who dare to dream.

Preface

This book highlights the background and fundamentals of the nonlinear optical materials in the perspective of the all-optical switching applications. Various second and third order nonlinear optical processes are attributed to the interaction of highly intense laser pulses with the materials. The interaction of the intense laser pulses with the materials is described in the first portion of the book which is followed by the introduction to the second and the third order nonlinear optical processes. Various topics such as origin of the nonlinear optical response, electronic response, slowly varying amplitude approximation for degenerate nonlinear refraction and nonlinear absorption, molecular reorientation response are elaborated at length. The nonlinear refractive index and the nonlinear absorption phenomenon are the intensity-dependent effects, which are placed in the category of the third-order nonlinear optical effects. These intensity dependent effects can be applied to fabricate different devices such as optical limiting, the all-optical switching, and detectors. The ratio of the nonlinear refractive index and the absorption coefficient values is known as figure-of-merit which is quite useful to decide whether a particular material is a good candidate for the all-optical switching. The spatial and temporal profiles of the ultrashort-pulsed lasers are elaborated in the second portion of the book by using the knife-edge and the autocorrelation techniques, respectively. Also the experimental techniques used to determine the nonlinear optical susceptibility value of the materials are discussed.

The third portion of the book deals with the measurement of the figure-of-merit of various organic molecules, which are useful for all-optical switching applications. The question "how to estimate the nonlinear absorption and refractive index of the organic molecules" is also addressed in the book. Various nonlinear organic molecules are explored to select the best materials in terms of the high figure-of-merit values. The question "how to engineer and design the organic molecules to produce high figure-of-merit values for the all-optical switching applications" is addressed. The last portion of the book deals with the various experimental schemes to realize the all-optical switching by using free space optics. A particular emphasis is given to a Mach-Zehnder interferometer switching symmetry where the optical and electronic components are described. I have also discussed how to optimize various parameters to obtain a phase change of Pi, which is attributed to a large value of the nonlinear refractive index of the organic dye.

There are issues associated with the fabrication of the all-optical switching devices. The transmission signal obtained in majority of the nonlinear optical effects is found to be weak which creates hiccups to obtain faster switching speeds. The various solutions are discussed to overcome the existing limitations of the all-optical switching based devices. The switch-on state of the all-optical switches is well defined and explored by many books, however, the switch-off state remains incomplete owing to the recombination time of the carriers which is discussed at a length in the current book. The Coherent Anti-Stokes Raman Scattering microscopy with details of the experimental configuration and applications are discussed. An introduction to the two-photon fluorescence and its applications to record images of the biological specimen. The discussion is followed by multiphoton microscopy and its applications. The nonlinear optical techniques in bioimaging and options for super-resolution microscopic imaging are also discussed.

Bilaspur, India Rajesh Sharma

Acknowledgements

The book *Nonlinear Optical Materials for All-Optical Switching Applications* has been a rewarding journey through the support and contributions of numerous individuals and organizations.

I would like to express my gratitude to Prof. Alok Kumar Chakrawal, Vice-Chancellor, Guru Ghasidas Vishwavidyalaya (A Central University), Bilaspur, Chhattisgarh, India, for motivation to perform high-quality research and publish in form of papers and books.

I am indebted to my colleagues and collaborators, particularly Dr. Manjot Singh who has completed his Ph.D. Degree under my supervision and has contributed by drafting text as well as new ideas incorporated in the present book. Thanks to the other students for their scientific ideas, Dr. Ayushi Chauhan, Dr. Reena Sharma, Dr. Deepak Kumar, and Dr. Harpreet Kaur, who has also completed their Ph.D. Degrees under my supervision. Their intellectual contributions and collaborative efforts have played a crucial role in shaping this research. Ms. Swati Das, Late Kaushtab Sadhu, and Ashish Kumar has also provided emotional and professional support. In cafeteria, we had several chat sessions on tea table with Dr. Awadhesh Kumar Dubey, Dr. Suchita, Dr. Gayatri Sarkar Guru Ghasidas Vishwavidyalaya (A Central University), Bilaspur, Chhattisgarh, India, which made the environment very light and full of fun.

Professor David Hagen, Dean, College of optics and photonics (CREOL), University of Central Florida, USA, who has motivated me to work in the field of ultrafast optics and switching. Thanks to Prof. Eric W. Van Stryland, Research Center in Electro-Optics and Lasers (CREOL), University of central Florida, Orlando, USA, for help when I worked as a Senior Research Scientist in his lab.

Thanks to Dr. Xianfan Xu, James J. and Carol L. Shuttleworth Professor in School of Mechanical Engineering Purdue University, USA for collaboration.

Special Thanks to Prof. Alexander Khmaladze, University at Albany, the State University of New York, USA, and Prof. Luca Varani, University of Montpellier, France, for close collaborations and motivation during my difficult time. Thanks to Dr. Ritu Shrivastava and Dr. Mahesh Kumar, ultrafast optics laboratory, National Physical lab in New Delhi, India, for pump-probe spectroscopy of the samples.

Thanks to Dr. K. P. Singh, Professor, Physical Sciences, Indian Institute of Science Education and Research, Mohali, India, and Dr. S. K. Mondal, Principal Scientist, Advanced Material and Sensors of the CSIR-Central Scientific Instruments Organisation, Chandigarh, India, for providing indirect support to my research group.

My sincere thanks to the editorial team at Springer Nature Publishing, especially my editor, Dr. Loyola D'Silva. His meticulous attention to detail and unwavering support throughout the publication process have been invaluable. Thanks to Dr. Vinothini Elango for her help at every stage of publication.

On a personal note, I am profoundly thankful to my family. My parents have been my pillars of strength, providing unwavering support and encouragement. God has blessed me with an intellectual mother whose standard of parenting can not be matched to date by any other person in my life. My spouse, Anita Sharma, and my children, Aashriya Sharma and Parnika Sharma, have been a constant source of inspiration and motivation, offering patience and understanding during this endeavor. My sister Neelam Sharma and her family also played an emotional role and encouragement throughout my career.

In conclusion, I extend my heartfelt thanks to all who have contributed to this book, whether through direct collaboration, financial support, or moral encouragement. Your collective efforts have made this work possible, and I am deeply grateful for your support. Overall, almighty GOD has given me everything as well as the required power and motivation to perform world-class research.

Contents

Abbreviations

ω	Angular frequency
n_2	Nonlinear refraction coefficient
α_2	Nonlinear absorption coefficient
χ^2	Second-order Susceptibility
χ^3	Third-order Susceptibility
1PA	One photon absorption
2PA	Two photon absorption
AOSP	All optical signal processing
CdSe	Cadmium selenide
DA	Dual arm
D-NLA	Degenerate nonlinear absorption
D-NLR	Degenerate nonlinear refraction
END-2PA	Extremely non-degenerate two photon absorption
END-NLR	Extremely non-degenerate nonlinear refraction
fs	Femtosecond
FWHM	Full width half maximum
GaAs	Gallium arsenide
ND-NLA	Non-degenerate nonlinear absorption
ND-NLR	Non-degenerate nonlinear refraction
NLA	Nonlinear absorption
NLO	Nonlinear optics
NLR	Nonlinear refraction
PP	Pump-probe spectroscopy
ps	Picosecond
TAS	Transient absorption spectroscopy
ZnS	Zinc sulphide
ZnSe	Zinc selenide

Chapter 1
Introduction to Nonlinear Optical Materials

1.1 Introduction to Nonlinear Optical Materials

Nonlinear optics (NLO) has received a lot of attention from researchers since the invention of lasers. NLO involves the interaction between light and matter, where an incident beam of light has sufficient intensity to change the optical properties of materials. Over the years, many nonlinear phenomena have been discovered experimentally, and theoretical models have been developed [2]. Many optical devices have been developed based on the NLO properties of materials, such as all-optical signal processing (AOSP) [10]. Researchers have made continuous efforts to find materials with large nonlinearity to improve the efficiency and performance of existing NLO devices. The second harmonic generation, third harmonic generation, sum frequency generation, difference frequency generation, optical switching [7, 15], four-wave mixing [17, 18], and optical limiting [4, 13, 20, 21] are among various topics which received significant attention of the researchers.

1.1.1 Linear and Nonlinear Optical Effects

The nonlinear optical properties of a material can be neglected if the incident electric field of the electromagnetic wave is small. On the other hand, when the incident electric field is very large, the nonlinearity of the material cannot be ignored. In such a situation, we enter the NLO realm, where many effects come into play such as second-order nonlinear effects [14], electro-optic effects, Pockels effects, second harmonic generation, sum frequency generation, difference frequency generation, down-conversion, parametric amplification [14], etc.

$$P = P_L + P_{NL} \tag{1.1}$$

R. Sharma, *Nonlinear Optical Materials for All-Optical Switching Applications*,
Progress in Optical Science and Photonics 33,
https://doi.org/10.1007/978-981-96-5921-0_1

The following equation represents nonlinear polarization which is sum of linear and nonlinear terms.

$$P(E) = \chi \varepsilon_0 E + \chi^{(2)} \varepsilon_0 E^2 + \chi^{(3)} \varepsilon_0 E^3 + \ldots \tag{1.2}$$

Here, the first term represents the linear polarization, while χ^2 and χ^3 terms correspond to the second-order and third-order nonlinear susceptibility, respectively. The polarization verses applied electric field is linear in a linear dielectric medium which is highlighted in Fig. 1.1a. The linear relationship between polarization and the electric field holds until the applied electric field is small, however, it becomes nonlinear when the applied electric field (Fig. 1.1b) is equivalent to the interatomic field i.e., of the order of 10^5 to 10^8 V/m. When the incident electric field is proportional to E^3, the third-order nonlinear term and the self-induced effect dominate. In the present book, we shall focus on the third-order nonlinear effect χ^3. These are totally intensity-dependent changes in the material properties, such as the self-focusing effect, self-phase and cross-phase modulation, two-photon absorption [21], the Raman effect, and the intensity-dependent refractive index of materials (nonlinear refractive index).

The NLO processes have their origins in the response to the high intensity of light, which affects the intra-atomic field of the materials to produce different nonlinear effects. In the case of the first-order term in equation (2), one can estimate that the material's response is proportional to the strength of the applied electromagnetic fields. One can derive the number of different optical processes, such as linear refraction and absorption, by using the first-order term. These linear phenomena are widely studied and well understood from theoretical as well as being implemented

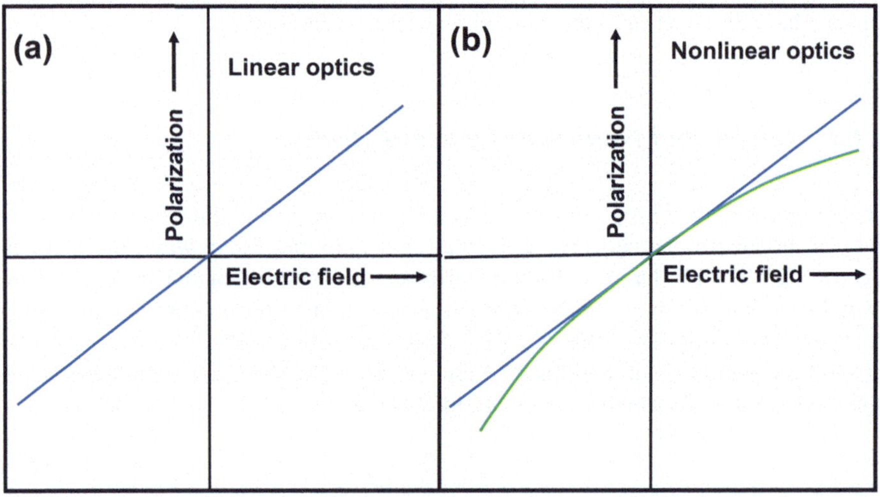

Fig. 1.1 Relation between polarization and electric field for **a** linear dielectric medium and **b** nonlinear dielecric medium

in many experimental optical components, such as lenses, fiber optics, etc. After the linear term, considering the higher-order terms in the response of the material under the influence of strong electromagnetic fields, one can derive different NLO phenomena, such as sum frequency generation (SHG), difference frequency generation (DFG), optical rectification (OR), nonlinear absorption (NLA), and nonlinear refraction (NLR) [1]. These phenomena are utilized to implement in many NLO devices, such as frequency converters, optical limiters, and optical switches [16]. However, the NLO response of many materials of interest is not well understood, which makes it an exciting area of research.

These aspects can be understood by the Lorentz model where restraining elastic force is proportional to the displacement and Hook's law is satisfied. On the other hand if the elastic force is increased beyond certain level, the linear relationship between the elastic force and displacement does not hold. In this example, the polarization and electric field are considered to the equivalent elastic force and displacement, respectively.

The χ^3 process occurs to some degree in all materials. Historically, the effect was first reported by John Kerr [23], who discovered materials that induced birefringence; the nonlinear refractive index (NLR) has a quadratic dependence on the electric field. The mechanism of χ^3 nonlinearity is attributed to distortion in the electron clouds surrounding the nucleus in an atom or molecule [14]. It polarizes atoms or molecules of the material owing to the applied strong electric field in the form of electromagnetic radiation. In 1875, John Kerr used a paraffin lamp, but nowadays researchers employ high-power laser sources to produce a strong electric field because χ^3 processes are electronic in origin. Recently, researchers have used ultrashort laser pulses to analyze the nonlinear optical properties of materials, owing to the fact that continuous wave lasers produce heat in the material, but the pulse fluence is high in the case of ultrashort lasers. All types of materials show third-order nonlinear optical effects irrespective of any inversion symmetry. Nonlinear polarization for χ^3 processes is proportional to the cube of the incident electric field. The nonlinear index of refraction associated phenomena of self-focusing and nonlinear absorption is discussed. To use a particular material for all-optical switching and optical limiting applications, the essential condition of obtaining high NLR values and low nonlinear absorption (NLA) values must be met. The NLR is a process where the refractive index of the material depends on the intensity of light rather than the wavelength.

$$n(I) = n_0 + n_2(I) \tag{1.3}$$

where n_0 and n_2 are the linear and nonlinear refractive indices, respectively. The NLR is used in various optics and photonics applications, such as mode-locking, optical switching etc.

The third-order effects have many applications. However, one application of particular interest to the scientific community is all-optical signal processing (AOSP) [11]. Currently, in the telecommunication system, the communication signal is transmitted via fiber optic technology although information switching and processing mainly depend on the electronic components. AOSP would offer significant advantages over the existing switches because the optical Kerr effect (OKE) comes into play due to the bound electronic nonlinearity, which makes it an ultrafast process occurring at the sub-femtosecond time scale [6]. The Kerr effect-based nonlinear process is faster than free electron transition and the thermal process.

Typically, the techniques for determining NLA and NLR include using a highly intense laser beam to induce nonlinear effects in the materials. The change in the material's optical properties induced by the intense beam is then measured with the help of different techniques such as optical Kerr measurement, four-wave mixing techniques, and pump-probe [12]. NLA is usually easier to measure using the above-mentioned techniques. Since NLA minimizes the transmission through the material, it can be easily determined by measuring the received energy of the probe beam using a photodiode. NLR, on the other hand, is relatively difficult to measure because photodiodes are not sensitive enough to the phase shift of the probe beam.

1.1.2 Second-Order Nonlinear Optical Effects

Second-order nonlinear optical effects refer to a class of phenomena that occur in materials when they are exposed to an intense electromagnetic field. Two beams of frequencies, ω_1 and ω_2, incident on the nonlinear crystal can mix to produce a third beam at a frequency of $\omega_1+\omega_2$; the effect is called sum frequency generation which is highlighted in the top panel of Fig. 1.2. These effects are characterized by the generation of new frequencies that are related to the original frequency of the input light through a process called second harmonic generation (SHG). SHG is a type of frequency doubling that occurs when two photons of the same frequency combine to produce a single photon of twice the frequency. This process is highly sensitive to the material properties and can be used to probe the electronic structure of molecules and materials. When the signal, ω_1, and pump beam, ω_3, are incident on the nonlinear crystal, the signal beam is amplified and a second beam at frequency ω_2 is produced which is known as the idler beam. The process, where the signal beam is

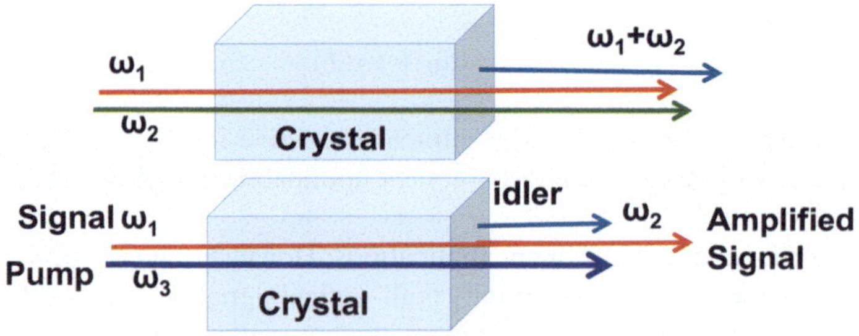

Fig. 1.2 Sum frequency generation and optical parametric amplification processes

amplified and idler beam is generated, is known as optical parametric amplification and is highlighted in bottom panel of Fig. 1.2. The optical parametric amplifier can be converted into optical parametric oscillator if proper feedback mechanism is used. One of the most widely used methods to study second-order nonlinear optical effects is through the use of pulsed lasers. Pulsed lasers generate high intensity light beams that can excite materials to their nonlinear regime.

The physical mechanisms underlying second-order nonlinear optical effects can be described in terms of a polarization response function, which relates the induced polarization of a material to the strength of the electric field that drives it. In second-order nonlinear materials, the polarization response function is nonlinear and can be expanded in a power series to include higher order terms.

The origins of the nonlinear response of materials can be traced back to the quantum mechanics of electronic transitions. When a material is exposed to light, the electrons in the material absorb energy and move to higher energy states. In a linear material, these transitions are allowed only for specific frequencies of light. In contrast, in a nonlinear material, the energy levels of the electrons are shifted due to the presence of an intense electric field, leading to new transitions that can give rise to SHG and other nonlinear effects. Second-order nonlinear optical effects have a wide range of applications in various fields such as communication, sensing, imaging, and spectroscopy. The development of nonlinear optical microscopes is an application, which use the nonlinear response of materials to obtain high resolution images of biological samples.

In recent years, there has been a growing interest in the development of new materials that exhibit strong second-order nonlinear optical effects. One approach is to use organic molecules, which have a high degree of symmetry and can exhibit strong nonlinear responses. Another approach is to use inorganic materials such as ferroelectric crystals, which have a large spontaneous polarization that can be oriented through an external electric field. The development of new materials with strong second-order nonlinear optical effects has the potential to revolutionize various fields such as quantum computing, photonics, and nonlinear optics.

The second-order nonlinear optical effects are a fascinating area of research that has important applications in various fields. The study of these effects involves the use of intense light beams and sophisticated measurement techniques to probe the nonlinear response of materials.

1.1.3 Third-Order Nonlinear Optical Effects

Third-order nonlinear optical effects refer to the phenomena in which the optical properties of a material are altered by the interaction of three or more photons. These effects play an important role in various applications such as all-optical switching, frequency conversion, optical signal processing, and optical communication. Here, we will discuss the different types of third-order nonlinear optical effects and their applications. The third-order optical susceptibility, $\chi^{(3)}$, is a measure of the third-order nonlinear optical response of a material. It relates the induced polarization of the material to the cubic power of the electric field. The third-order susceptibility can be expressed as a tensor quantity due to the anisotropic nature of the material.

One of the most important third-order nonlinear optical effects is the third harmonic generation (THG), which involves the generation of a new photon at three times the frequency of the incident photon. THG can be used to generate ultraviolet radiation from visible or near-infrared light sources. This effect is commonly observed in materials with a large $\chi^{(3)}$ such as semiconductors, metals, and glasses. Another important third-order nonlinear optical effect is the nonlinear refraction (NLR), which occurs due to the intensity-dependent change in the refractive index of a material. NLR can lead to self-focusing, self-phase modulation, and other nonlinear optical phenomena. NLR is usually observed in materials with a large electronic polarizability and a fast response time.

In addition to THG and NLR, there are other third-order nonlinear optical effects such as four-wave mixing (FWM), stimulated Raman scattering (SRS), and stimulated Brillouin scattering (SBS). FWM involves the generation of two new photons at frequencies that are the sum and difference of the frequencies of the incident photons. SRS and SBS involve the interaction of photons with phonons in the material, leading to the generation of new photons at frequencies that are shifted from the incident photons. Third-order nonlinear optical effects have numerous applications in various fields. For example, THG can be used for frequency conversion in optical communication systems. NLR can be used for all-optical switching and optical limiting. FWM can be used for wavelength conversion in optical communication systems. SRS and SBS can be used for sensing and measurement applications. One of the major challenges in the development of third-order nonlinear optical materials is to achieve a large $\chi^{(3)}$ with a fast response time. This requires the synthesis and design of novel materials with specific electronic and structural properties. Various approaches have been explored to enhance the $\chi^{(3)}$ of materials, including the use of conjugated organic molecules, metal nanoparticles, and semiconductor quantum dots.

In conclusion, third-order nonlinear optical effects play a crucial role in various applications such as all-optical switching, frequency conversion, and optical signal processing. The development of novel materials with enhanced $\chi^{(3)}$ and fast response time is essential for the advancement of these applications. The study of third-order nonlinear optical effects continues to be an active area of research, and new materials and applications are constantly being explored.

1.1.4 Nonlinear Absorption and Refraction

Nonlinear optical effects are the phenomena that arise when light interacts with materials and the response of the material is no longer linearly proportional to the intensity of the light. Nonlinear absorption and refraction are two important third-order nonlinear optical effects that have significant practical applications in a range of fields such as telecommunications, optical switching, and laser technology. Nonlinear absorption occurs when the absorption of light by a material depends nonlinearly on the intensity of the light. This means that the rate of absorption of light increases as the intensity of the light increases beyond a certain threshold. This can be explained by considering the energy levels of the atoms or molecules in the material. In a linear absorption process, the incident photon promotes an electron from its ground state to a higher energy state, which then decays back to the ground state by emitting a photon of the same energy. However, in a nonlinear absorption process, the energy of the absorbed photon is used to excite multiple electrons to higher energy levels, leading to a higher rate of absorption. NLA is one of the processes that arise due to χ^3 nonlinearity. In this process, multiple photons are absorbed to excite the electron from the ground state to the excited state. The process is depicted in Fig. 1.3.

Nonlinear absorption can have both detrimental and beneficial effects depending on the application. In telecommunications, for example, it can cause signal distortion and loss, leading to reduced transmission distances and data rates. However, in optical limiting and protection devices, nonlinear absorption is desirable as it can effectively protect sensitive optical components by limiting the amount of incident light that reaches them.

Nonlinear refraction, on the other hand, occurs when the refractive index of a material depends nonlinearly on the intensity of the light. This means that the speed of light in the material changes nonlinearly with the intensity of the light, leading to a change in the direction of the light wave. This can be explained by considering the response of the electrons in the material to the electric field of the light. In a linear refraction process, the electric field causes a displacement of the electrons from their equilibrium positions, leading to a change in the refractive index of the material. However, in a nonlinear refraction process, the displacement of the electrons is larger and more complex, leading to a nonlinear dependence of the refractive index on the intensity of the light. Nonlinear refraction can also have both detrimental and beneficial effects depending on the application. In telecommunications, it can cause signal distortion and dispersion, leading to reduced transmission distances and data rates. However, in optical switching and modulation devices, nonlinear refraction

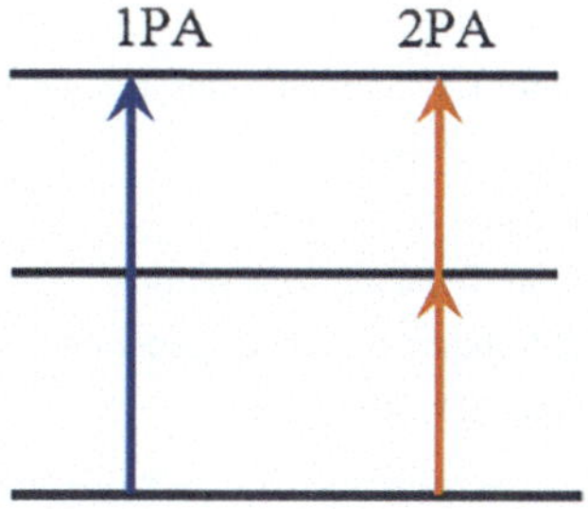

Fig. 1.3 Schematic diagram of one (left arrow) and two (right arrows) photon absorption mechanism

is desirable as it can be used to control the transmission of light by changing the refractive index of the material.

It is important to note that nonlinear absorption and refraction are often interrelated and can occur simultaneously in a material. This means that the response of the material to light is complex and nonlinear, and can vary depending on the intensity, wavelength, and duration of the incident light. Additionally, the nonlinear response of a material can be affected by its physical and chemical properties, such as its composition, crystal structure, and temperature. To measure the nonlinear absorption and refraction of a material, various experimental techniques are available, such as the Z-scan technique, the pump-probe technique, and the transient grating technique. These techniques involve measuring the transmission and/or reflection of light through the material as a function of the incident intensity and/or time delay.

In conclusion, nonlinear absorption and refraction are important third-order nonlinear optical effects that have significant practical applications in various fields. They can have both detrimental and beneficial effects depending on the application, and their interrelated nature makes the response of materials to light complex and nonlinear. Experimental techniques are available to measure the nonlinear response of materials, which can help in designing and optimizing nonlinear optical devices and systems.

1.2 Origin of Nonlinear Optical Response

The optical response of materials to incident radiation is schematically highlighted in this section. The effect of varying electric field amplitude on polarization is explained by taking into account the second and third-order optical nonlinearity of the materials. The second and third-order optical nonlinear mechanisms are also discussed using the input and output frequencies in Table 1.1 [1].

At the molecular scale, the Raman response can be understood in terms of the coupling of vibrational and electronic energy states [2]. Molecular reorientation occurs when molecules are reoriented in the presence of an incident polarized electromagnetic field [16]. Additionally, inter-molecular interaction occurs when molecules collide with each other [11]. The molecular liberation is also taken into account to

Table 1.1 Second-order nonlinear mechanisms

Mechanisms	Input frequency	Output frequency
Frequency doubling	ω	2ω
Sum frequency generation	ω_1, ω_2	$\omega_3 = \omega_1 + \omega_2$
Difference frequency generation	ω_1, ω_2	$\omega_3 = \omega_1 - \omega_2$
Optical rectification	ω	DC electric field

define the inter-molecular interaction because of the oscillation of the molecules in the presence of the field of nearby molecules [11]. Moreover, in dipole-dipole interactions, the dipole of one molecule may affect the dipole of a neighboring one [19].

Recently, Zhao et al. [24] measured the NLO response of 24 different solvents with high accuracy using the beam-deflection technique. The beam-deflection technique allows determination of the electronic and nuclear contribution of the nonlinearity separately. Four types of nonlinearities are mainly present in organic molecules and gases: bound-electronic, nuclear collision, liberation, and diffusive reorientation. The nonlinearity due to the bound electrons is treated as an instantaneous nonlinearity, which is faster than the nuclear contribution [24]. The optically induced molecular reorientation or diffusive reorientation can produce changes in the NLR values of the NLO organic materials [3, 9, 22]. The anisotropic molecule is polarized when the incident optical field makes an angle with the induced dipole moment; the phenomenon is called diffusive reorientation. The liberation and diffusive reorientation have similar origins and symmetries, however, the field-induced dipole re-radiates the electric field which changes NLR values via distortion of the polarizability in nearby molecules.

For instance, in 2015, Iliopoulos et al. reported the third-order NLO properties of commonly used organic solvents, which were obtained by ps and fs laser pulse excitation. The Z-scan measurements were recorded at 35 ps and 40 fs for the wavelength values of 532 and 800 nm, respectively. The OKE study was performed at 35 ps at the wavelength of 532 nm [8].

Acoustic waves can alter the refractive index of the molecules of the liquid owing to the density variation at a macroscopic scale. The pulsed laser beam can be directed onto the piezoelectric crystal to produce the acoustic wave signal. Thermal optical nonlinearities arise owing to the production of heat in the medium by the pulsed laser [16]. Typically, several of the factors discussed above are responsible for the nonlinear nature of the materials under investigation. In the present book, we are interested in the Kerr nonlinearity, which occurs at an ultrafast timescale. The electronic, nuclear, and reorientational responses are explored to identify the nonlinear nature of the materials.

1.2.1 Electronic Response

An oscillatory electric field can induce the optical changes in the material

$$E = E_0 \frac{1}{2}(e^{-i(kz-\omega t)} + e^{i(kz-\omega t)}) \tag{1.4}$$

where E_0, k, z, ω and t are the electric field strength, wave-number, the propagation distance, angular frequency ($\omega = 2\pi \upsilon$) and time, respectively. The time-varying electric field exerts a force to induce oscillatory motion of the electron clouds and nucleus. Since the relative mass of the nucleus is higher than the electron cloud, the induced motion of the nucleus is negligible compared to the electron cloud, so one can ignore the motion of the nucleus and limit it to the distortion of the electron cloud. The distortion of the electron cloud induces the dipole moment in the atom. The dipole moment of the atom can be written and expanded using the power series.

$$\mu_i = \mu_i{}^0 + \varepsilon_0(\alpha_{ij}E_j + \beta_{ijk}E_iE_j + \gamma_{ijkl}E_jE_jE_l + ...) \tag{1.5}$$

where $\mu_i{}^0$ is the permanent dipole moment, α_{ij} is the first order polarizability, β_{ijk} and γ_{ijkl} are the second and third-order polarizabilities, respectively. The Eq. 5 can be rewritten in terms of the macroscopic polarization,

$$P_i = p_i{}^0 + \varepsilon_0\chi_i{}^{(1)}E_j + \chi_{ij}^{(2)}E_jE_k + \chi_{ijk}^{(3)}E_jE_kE_l + ...\chi_{ijk...n}^{(n)}?E_i...E_n) \tag{1.6}$$

The first term, p_i^0, is the permanent polarization moment, the second term, $\varepsilon_0\chi_i^{(1)}E_j$, is the linear polarization response to the driving field, which accounts for linear properties of the material such as the linear refractive index and the linear absorption phenomena. The higher-order terms are responsible for the nonlinear properties of the material such as second harmonic generation, third harmonic generation, sum-frequency mixing, difference frequency mixing, NLR, and NLA, etc.

To obtain the first-order polarization response, the time-varying electric field can exert the restoring force, which is proportional to the displacement of the electronic cloud from the mean position. The second term in the expansion of Eq. 1.6 is $\mu = \alpha E$, where α is the polarization of the material. The bulk polarization, P, for atomic density N, is given as follows:

$$P = N\mu = N\alpha E = \varepsilon_0\chi^{(1)}E \tag{1.7}$$

The parameters, N and α, allow us to characterize many of the linear optical phenomena, such as the absorption, reflection, refraction and transmission etc. However, still there are many processes that cannot be adequately explained with the first order theory, especially when dealing with the high irradiance values of the laser beam.

For the second-order nonlinear response, the polarization response is

$$P^{(2)}\omega = \varepsilon_0 \int_{-\infty}^{\infty}\int_{-\infty}^{\infty} \chi^{(2)}(\omega, \omega_a, \omega_b)E(\omega_a)E(\omega_b) + \delta(\omega, -\omega_a, -\omega_b)d\omega_a d\omega_b \tag{1.8}$$

Equation 1.8 represents the two EM-waves at two different frequencies, $E(\omega_a)$ and $E(\omega_b)$, which results in 16 terms that can be grouped together in classes of similar effects. There is the polarization response $P^{(2)}(2\omega_a)$, $P^{(2)}(2\omega_b)$ which are referred to as second harmonic generation (SHG). The SHG can be observed in various nonlinear optical materials such as potassium dihydrogen phosphate (KDP), $\beta-$barium borate (BBO), lithium triborate (LBO) which can be used in laser systems to double the incident frequency.

Table 1.2 Third-order nonlinear mechanisms

Third-order mechanisms	Input frequency	Output frequency
Frequency tripling	ω	3ω
Self-phase modulation or Degenerate nonlinear refraction	ω	Phase shifted for frequency ω
Degenerate two-photon absorption	ω	ω Attenuated
Degenerate four-wave mixing	Three beam at ω	New beam at ω
Cross phase modulation or nonlinear refraction (Nondegenerate)	ω_1,ω_2	ω_1, ω_2, phase shifted
Two-photon absorption (Nondegenerate)	ω_1, ω_2	ω_1, ω_2, attenuated
Four wave mixing (Nondegenerate)	ω_1, ω_2, ω_3	ω_4

The third-order nonlinearity arises from the third-order polarization term in eqn. 1.6. The third-order term can be written as

$$P^{(3)}\omega = \varepsilon_0 \int_{-\infty}^{\infty}\int_{-\infty}^{\infty} \chi^{(3)}(\omega, \omega_a, \omega_b, \omega_c)E(\omega_a)E(\omega_b)E(\omega_c)x\delta(\omega, -\omega_a, -\omega_b, -\omega_c)d\omega_a d\omega_b d\omega_c \tag{1.9}$$

The details of the various second-order nonlinear optical processes are highlighted in Table 1.1, where the input frequency, output frequency, and the term responsible for defining the parameters are schematically tabulated. In the case of SHG, the input frequency ω is converted to the frequency 2ω. On the other hand, the input frequencies ω_1 and ω_2 are converted into the laser beam with a frequency of $\omega_1 + \omega_2$, which is known as sum-frequency generation (SFG). In fact, SHG can be considered a special case of SFG where the incident beams are of equal frequencies.

Nevertheless, the output frequency of the laser beam $\omega_1 - \omega_2$ in the case of difference frequency mixing. However, optical rectification comes into the picture when the difference-frequency mixing occurs with equal input frequencies. Consequently, a static electric field is generated during the process of optical rectification. Table 1.2 summarizes the third-order optical nonlinear mechanisms reported by Fu [5]. Degenerate two-photon absorption (D-2PA), NLR, THG, and degenerate four wave mixing (D-FWM) are degenerate processes that can be obtained by using a single beam, while the other nondegenerate processes require an additional beam to obtain desired output frequencies.

1.2.2 Slowly Varying Amplitude Approximation

For the case of degenerate beam processes, the electric field is

$$E(z,t) = \frac{1}{2}E(z)(exp(i(kz + \omega t))) + c.c. \tag{1.10}$$

The wave equation for the propagation is

$$\nabla^2 E(z,t) - \frac{1}{c^2}\frac{\partial^2 E(z,t)}{\partial t^2} = \mu_0 \frac{\partial^2 P^{(3)}(z,t)}{\partial t^2} \tag{1.11}$$

For the case of nondegenerate beams, electric field is given as

$$E(z,t) = E_a(z,t) + E_b(z,t) \tag{1.12}$$

Generally, there will be a polarization response at a number of different frequencies as described in Table 1.2. In the present book, the main aim would be to investigate the NLR and NLA, which occur at frequency ω.

1.2.3 Molecular Reorientation Response

In the case of organic molecules, the instantaneous and non-instantaneous mechanisms are mainly responsible for changes in the NLR. The instantaneous mechanism is attributed to the bound electronic response, whilst in the case of the non-instantaneous mechanism, the nuclear motion, collision, liberation, and diffusive reorientation are responsible for changes in the NLR. The reorientation response comes into play when the molecules are rotated due to the excitation electric field associated with the laser beam.

Nowadays, optical fiber and other optical transmission lines are widely used in various communication systems. If the irradiance of the incident signal is high, nonlinear effects such as self-focusing and nonlinear absorption can be observed, leading to an increase in signal losses in the transmission medium due to nonlinearity. In this case, the NLR and NLA values of the materials are essential in minimizing the losses by choosing a particular material for the specific frequency range of signals. Researchers have used many techniques to measure the nonlinear coefficient of materials. The NLR and NLA have advantages for all-optical switching and optical limiting devices, which will be discussed in Sect. 4.1. The Z-scan is an efficient technique to measure the NLR and NLA values of the material.

1.3 Overview of Content of Book

Chapter 1 covers the background and fundamentals of the nonlinear optical materials in the perspective of the all-optical switching applications. The interaction of the intense laser pulses with the materials is described which is followed by the introduction to the second and the third-order nonlinear optical processes. The contents of the Chap. 1 can be easily found in the other standard books of lasers and nonlinear optics although a special emphasis is given to the applications of the third-order nonlinear susceptibility in the field of all-optical switching. Various topics such as origin of the nonlinear optical response, electronic response, slowly varying amplitude approximation for degenerate nonlinear refraction and nonlinear absorption, molecular reorientation response are elaborated at length.

In order to estimate the nonlinearity of a particular material, one must know the linear optical properties of the material such as linear absorption coefficient and linear index of refraction because the one-photon absorption energy levels can hinder the progress of the nonlinear absorption. Chapter 2 starts with an introduction to ultrafast laser beam configuration which is followed by several topics such as spatial and temporal profile estimation of the femtosecond pulsed laser systems, introduction to autocorrelation technique and estimation of pulse-width of ultrafast lasers etc. Chapter 3 deals advances in temporal profile measurement of ultrafast pulsed lasers. The various components of the autocorrelation configuration are elaborated at length. Nevertheless, frequency-resolved optical gating (FROG), spectral phase interferometry for direct electric-field reconstruction (SPIDER), grating-eliminated no-nonsense observation of ultrafast incident laser light e-fields (GRENOUILLE) are discussed to measure the temporal and spectral profiles of the ultrafast pulsed lasers. The laser beam spot size at different positions from the focus is described by using a knife-edge technique. Chapter 4 includes the topics such as advanced nonlinear optical techniques, the selection criteria of the nonlinear optical materials for the switching applications, single- and dual-arm Z-scan, pump-probe and beam-deflection techniques.

Chapter 5 introduces the degenerate two-photon absorption, the nondegenerate two-photon absorption and the extremely nondegenerate two-photon absorption mechanism in the semiconductor materials. Nevertheless, the extremely nondegenerate two-photon absorption coefficient values of selected semiconductor materials are significantly increased in comparison to the degenerate two-photon absorption and the nondegenerate two-photon absorption coefficient values. A special emphasis is given to explore the nonlinear absorption coefficient measurements of GaAs where one-photon absorption band is located around 873 nm whilst the two-photon absorption wavelength are obtained around 436nm. The possible applications of GaAs in the optical detectors are also explored. Chapter 6 deals spin-based dichroism and all-optical switching in nanocrystal quantum dots. In Chap. 6, various topics such as polarization dependence of degenerate pump-probe, noise optimization of all-optical switching signal, and solutions to various problems in all-optical switching in nanocrystal quantum dots and future direction are discussed. Chapter 7 deals with the estimation of the figure-of-merit of various organic molecules, which are useful for all-optical switching applications. The question "how to estimate the nonlinear absorption and refractive index of the organic molecules" is also addressed in this Chapter. Chapter starts with the investigations of the third-order nonlinear refractive cross section of the CS2 molecules. Various nonlinear organic molecules are explored to select the best materials in terms of the high figure-of-merit values. The question "how to engineer and design the organic molecules to produce high figure-of-merit values for the all-optical switching applications" is addressed in Chap. 7. Chapter 8 deals with all-optical switching in high figure-of-merit materials where various topis suc as examples of high figure-of-merit materials for all-optical switching, properties and characteristics of high figure-of-merit materials are elaborated at length. Chapter 8 also deals with various topics such as comparison of different all-optical switching mechanisms, experimental techniques for studying all-optical switching, examples of recent experimental studies on all-optical switching in high figure-of-merit materials etc. Chapter 9 elaborates challenges facing the development and implementation of all-optical switching, possible solutions to overcome these challenges and future directions of research in all-optical switching in high figure-of-merit materials

1.4 Summary

The origin of the different kind of nonlinearities such as electronic, molecular reorientation are discussed. The nuclear contribution can be exploited for different nonlinear optical applications with respect to their temporal dynamics. Nowadays, the optical fiber and other optical transmission lines are used in various communication systems. If the irradiance of the incident signal is high, the nonlinear effects such as self-focusing and nonlinear absorption can be observed. The losses of signal in transmission medium increase due to the nonlinearity.

References

1. R.W. Boyd, A.L. Gaeta, E. Giese, Nonlinear optics, in *Springer Handbook of Atomic, Molecular, and Optical Physics* (Springer, 2008), pp. 1097–1110
2. D.N. Christodoulides, I.C. Khoo, G.J. Salamo, G.I. Stegeman, E.W. Van Stryland, Nonlinear refraction and absorption: mechanisms and magnitudes. Adv. Opt. Photon. **2**(1), 60–200 (2010)
3. A. Dutt, A. Mohanty, A.L. Gaeta, M. Lipson, Nonlinear and quantum photonics using integrated optical materials. Nat. Rev. Mater. 1–26 (2024)
4. D.A. Fishman, C.M. Cirloganu, S. Webster, L.A. Padilha, M. Monroe, D.J. Hagan, E.W. Van Stryland, Sensitive mid-infrared detection in wide-bandgap semiconductors using extreme non-degenerate two-photon absorption. Nat. Photon. **5**(9), 561–565 (2011)
5. J. Fu, Molecular structure-nonlinear optical property relationships for a series of polymethine and squaraine molecules (2006)
6. S.A. Haque, J. Nelson, Toward organic all-optical switching. Science **327**(5972), 1466–1467 (2010)

7. S.M. Hendrickson, C.N. Weiler, R.M. Camacho, P.T. Rakich, A.I. Young, M.J. Shaw, T.B. Pittman, J.D. Franson, B.C. Jacobs, All-optical-switching demonstration using two-photon absorption and the zeno effect. Phys. Rev. A-Atom. Molecul Opt. Phys. **87**(2), 023808 (2013)
8. K. Iliopoulos, D. Potamianos, E. Kakkava, P. Aloukos, I. Orfanos, S. Couris, Ultrafast third order nonlinearities of organic solvents. Opt. Exp. **23**(19), 24171–24176 (2015)
9. B. Kim, J. Jin, Z. Wang, L. He, T. Christensen, E.J. Mele, B. Zhen, Three-dimensional nonlinear optical materials from twisted two-dimensional van der Waals interfaces. Nat. Photon. **18**(1), 91–98 (2024)
10. J. Leuthold, C. Koos, W. Freude, Nonlinear silicon photonics. Nat. Photon. **4**(8), 535–544 (2010)
11. D. McMorrow, W.T. Lotshaw, G.A. Kenney-Wallace, Femtosecond optical kerr studies on the origin of the nonlinear responses in simple liquids. IEEE J. Quantum Electron. **24**(2), 443–454 (1988)
12. L.A. Padilha, A.A.R. Neves, E. Rodriguez, C.L. Cesar, L.C. Barbosa, C.H. Cruz, Ultrafast optical switching with CDTE nanocrystals in a glass matrix. Appl. Phys. Lett. **86**(16) (2005)
13. H.S. Pattanaik, M. Reichert, J.B. Khurgin, D.J. Hagan, E.W. Van Stryland, Enhancement of two-photon absorption in quantum wells for extremely nondegenerate photon pairs. IEEE J. Quantum Electron. **52**(3), 1–14 (2016)
14. P.E. Powers, J.W. Haus, *Fundamentals of Nonlinear Optics* (CRC press, 2017)
15. A. Rostami, Low threshold and tunable all-optical switch using two-photon absorption in array of nonlinear ring resonators coupled to mzi. Microelectron. J. **37**(9), 976–981 (2006)
16. G.I. Stegeman, R.A. Stegeman, *Nonlinear Optics: Phenomena, Materials and Devices*, vol. 78 (Wiley, 2012)
17. R. Stolen, Phase-matched-stimulated four-photon mixing in silica-fiber waveguides. IEEE J. Quantum Electron. **11**(3), 100–103 (1975)
18. R. Stolen, J. Bjorkholm, Parametric amplification and frequency conversion in optical fibers. IEEE J. Quantum Electron. **18**(7), 1062–1072 (1982)
19. B.T. Thole, Molecular polarizabilities calculated with a modified dipole interaction. Chem. Phys. **59**(3), 341–350 (1981)
20. E.W. Van Stryland, H. Vanherzeele, M.A. Woodall, M.J. Soileau, A.L. Smirl, S. Guha, T.F. Boggess, Two photon absorption, nonlinear refraction, and optical limiting in semiconductors. Opt. Eng. **24**(4), 613–623 (1985)
21. E.W. Van Stryland, Y.-Y. Wu, D.J. Hagan, M.J. Soileau, K. Mansour, Optical limiting with semiconductors. JOSA B **5**(9), 1980–1988 (1988)
22. Z. Wang, Y. Yan, J. Chen, Q.-H. Li, J. Zhang, Designed metal-organic π-clusters combining the aromaticity of the metal cluster and ligands for a third-order nonlinear optical response. Mater. Horizons **11**(1), 297–302 (2024)
23. P. Weinberger, John Kerr and his effects found in 1877 and 1878. Philosoph. Mag. Lett. **88**(12), 897–907 (2008)
24. P. Zhao, M. Reichert, S. Benis, D.J. Hagan, E.W. Van Stryland, Temporal and polarization dependence of the nonlinear optical response of solvents. Optica **5**(5), 583–594 (2018)

Chapter 2
Spatial and Temporal Profile Measurements of the Lasers

2.1 Spatial Profile Measurement of Laser Beam

2.1.1 Ultrafast Laser Beam Configuration

In present work, the primary nonlinear optical properties of interest are NLA and NLR. The nonlinear characterization of these materials were carried out by Ultrafast laser; a Ti-Sapphire femtosecond laser system at 800 nm wavelength. The pulse width, repetition rate and the energy per pulse values of the fs laser system are 35 fs, 1KHz and 4 mJ respectively. The beam is directed onto an optical parametric amplifier (OPA) used to produce required excitation wavelengths. Then this pulsed laser beam is utilized for different nonlinear characterizations.

2.1.2 Experimental Details Knife Edge Technique

The knife-edge technique is a laser beam profiling technique that permits accurate, quick and inexpensive determination of parameters of the laser beam. It is a principle method for the characterization of gaussian beam profile. This method involves translation of knife-edge perpendicular to the direction of propagation of the laser beam. The total transmitted power is calculated as a function of position of the knife-edge which is mounted over the millimeter scale.

A He–Ne laser at 632.8 nm wavelength is used to perform the knife edge experiment. The optical alignment of laser beam is done so that the beam propagates in a straight line, as shown in Fig. 2.1. The entire experimental setup is explained as follows; the laser beam passes through the polarizer by which can attenuate the laser power and then the beam is directed onto the convex lens to focus down at a particular distance. In order to optically align the experimental configuration, the beam must travel along optical axis of all of the components [3]. After wards, the knife screw

R. Sharma, *Nonlinear Optical Materials for All-Optical Switching Applications*,
Progress in Optical Science and Photonics 33,
https://doi.org/10.1007/978-981-96-5921-0_2

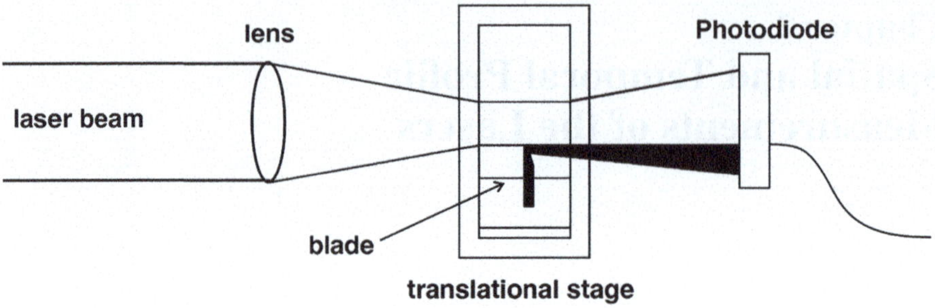

Fig. 2.1 Experimental setup to find out the radius of laser beam with knife-edge technique. The shaded area is representing the reflection made by the edge of knife [3, 16]

is moved to block the beam partially and then the transmitted intensity is detected by the photo diode. The detector is adjusted at maximum power of the transmitted beam [16]. Then the knife-edge is translated towards and away from the focus and the transmission signal is measured along y-axis and z-axis. This exercise will be used to estimate M^2 value of the beam profile. In the present experiment, two different convex lens of focal length 10 cm and 20 cm are used. The value of beam radius is calculated by using "user defined function" in origin lab program by performing the nonlinear curve fitting.

The essential components of the experimental setup are; He–Ne laser, polarizer, convex lens, sharp edge typically a razor blade, photo-detector and laser Power meter.

2.1.3 *Intensity Versus Position of Knife Edge*

The knife-edge technique is used to investigate the spatial profile of the laser beam where a convex lens of focal length 10 cm is used to focus the beam. The knife user

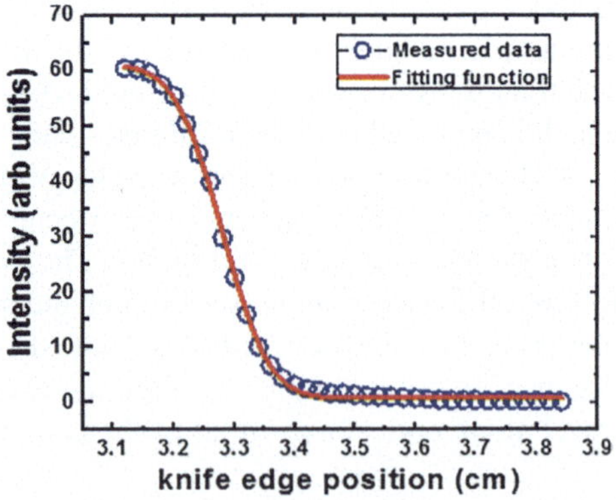

Fig. 2.2 Intensity as a function of position of the knife edge at a distance 4.4 cm towards the lens. The open circle symbols are the measured data points while the line is the fitting function obtained by using Eq. 2.1. The estimated value of the beam radius is 124 μm

function of origin lab program is used for the nonlinear curve fitting of the measured data (Fig. 2.2). The beam radius is found to be 124 μm. The position of the knife edge is at a distance 4.4 mm towards the lens from the focus, and the knife edge fitting function is given by

$$y = y_0 - A_0(erf\frac{\sqrt{2}}{r})(x - x_0) \tag{2.1}$$

where, erf is error function defined as

$$erf(x) = \frac{r}{\sqrt{\pi}} \int_0^x e^{-S^2} ds \tag{2.2}$$

In Figs. 2.3 and 2.4 the values of beam radius is found to be 192 and 218 μm by using nonlinear curve fitting and the knife-edge function. The position of knife edge is at a distance 3.5 and 2.5 cm towards the lens and the value of radius calculated by using knife edge fitting function, Eq. 2.1. The value of radius is calculated at different distances from the focus by using the same procedure.

2.1.4 M^2 Profile

The mode of the TEM_{00} laser beam propagation is determined by the M^2 value which in turn can be used to measures size of beam waist. Figure 2.5 show M^2 values of -4.468 obtained for the lens with the focal length of 10 cm respectively. In order to estimate the perfect gaussian symmetry the value of (M^2) should be equals to 1. The Gaussian beam profile is determined by M^2 calculation from the values of radii measured at different positions of the knife-edge.

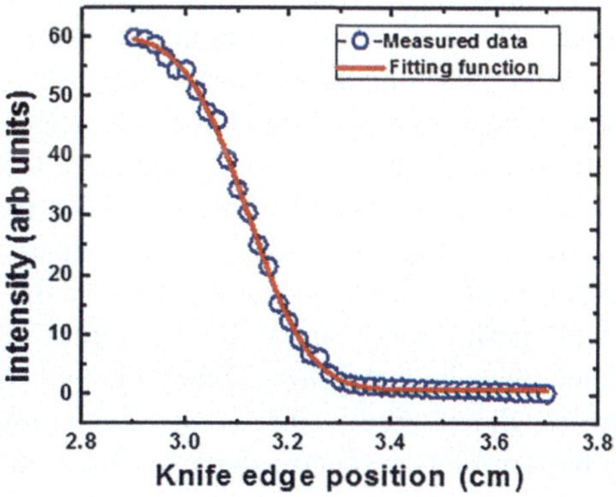

Fig. 2.3 Intensity as a function of position of knife edge at a distance 3.5 cm towards the lens. The open circle symbols are the measured data points while the line is the fitting function obtained by using Eq. 2.1. The estimated value of the beam radius is 192 μm

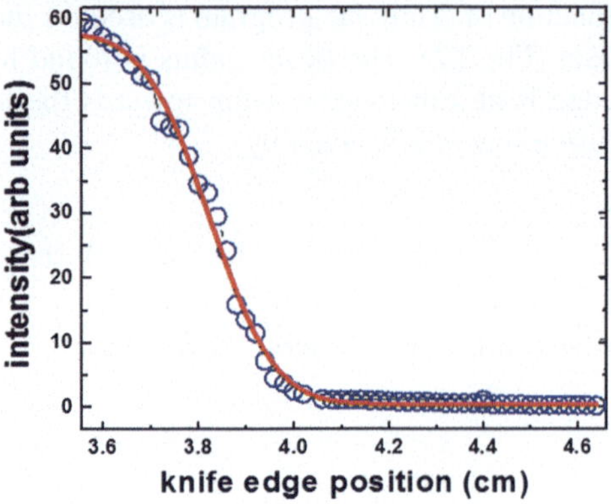

Fig. 2.4 Intensity as a function of position of knife edge at a distance 2.5 cm towards the lens. The open circle symbols are the measured data points while the line is the fitting function obtained by using Eq. 2.1. The estimated value of the beam radius is 218 μm

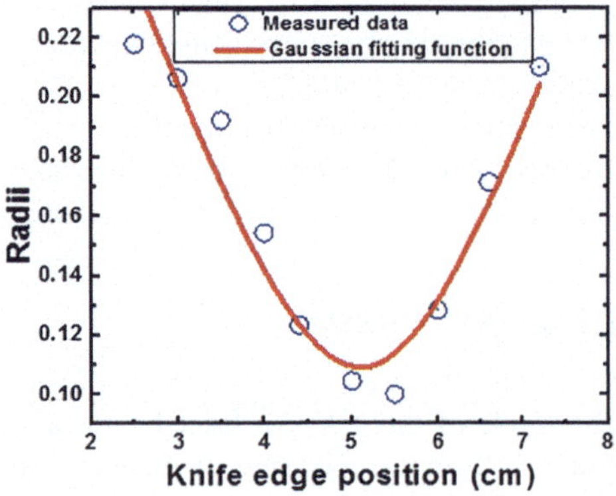

Fig. 2.5 The beam radii as a function of the knife edge position obtained by using plots and various knife edge positions along beam propagation direction (Z-axis) using lens of focal length 10 cm. The value of M^2 is calculated from Gaussian propagation fitting function

2.1.5 Spatial Profile Measurement of Laser Beam

In order to probe the spectroscopic characteristics of the nonlinear optical materials it is important to know about the laser spatial profile in continuous as well as pulse mode. There are many techniques to deduce spatial profile of laser.

After 1960s the lasers were widely used in various fields such as optics, nonlinear optics, laser cutting, biomedical sciences, micro-machining etc. With these applications the demand of laser source and technology is enhanced which is essential to determine the laser parameters like wavelength, intensity, beam spot size, beam profile with high accuracy. This information is essentially required to perform the experiment with high accuracy because the input parameters are responsible for the stable performance of the laser. In order to obtain stable performance of the lasers researchers have developed many methods to determine the beam profile of laser [3]. Although techniques such as ray tracing, multi-photon ionization yields and fluorescence correlation spectroscopy predict accurate results. These methods are either too complicated for most general applications or they rely on specific experimental procedures not easily adaptable to other experiments. In present section we discuss the knife edge technique to measure the laser beam profile using a home-made experimental setup and compare the experimental data with knife edge fitting function to extract the beam radius of Gaussian beam with different focal length lenses at different Z-positions and further we determine the beam radius at many point on Z-axis along the beam propagation direction. At last we determined the Gaussian profile of the beam using Gaussian fitting function or M^2 function which can tell us the about the Gaussian profile of the laser. If value of M^2 is 1 it is perfect Gaussian shape, however, the value other than 1 indicates deviation from the Gaussian beam (Fig. 2.6).

The position of the knife edge is at a distance 4.4 mm towards the lens from the focus, and the knife edge fitting function is given by Eq. 2.1 and error function is given by Eq. 2.2. In Fig. 2.7 the value of beam radius is found to be 192 μm by using nonlinear curve fitting and the knife-edge function. The position of knife edge is at a distance 3.5 mm towards the lens and the value of radius calculated by using knife edge fitting function, Eq. 2.1. The value of radius is calculated at different distances from the focus by using the same procedure (Fig. 2.8).

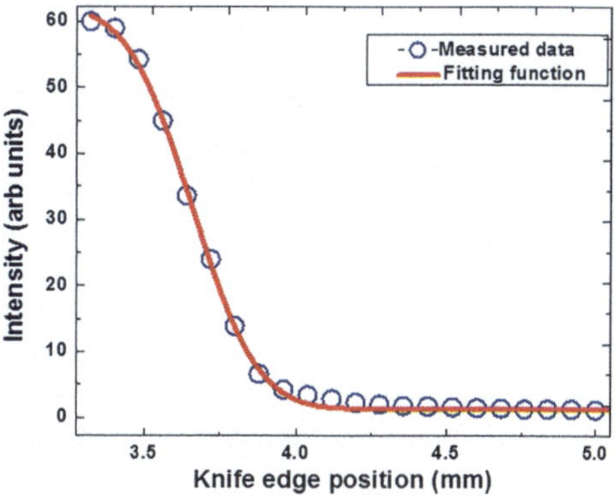

Fig. 2.6 Intensity as a function of position of the knife edge at a distance 10.5 cm towards the lens. The open circle symbols are the measured data points while the line is the fitting function obtained by using Eq. 2.1. The estimated value of the beam radius is 341 μm

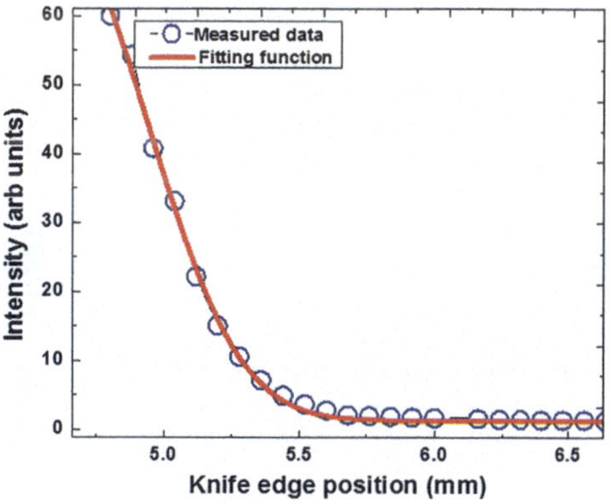

Fig. 2.7 Intensity as a function of position of knife edge at a distance 6.5 cm towards the lens. The open circle symbols are the measured data points while the line is the fitting function obtained by using Eq. 2.1. The estimated value of the beam radius is 541 μm

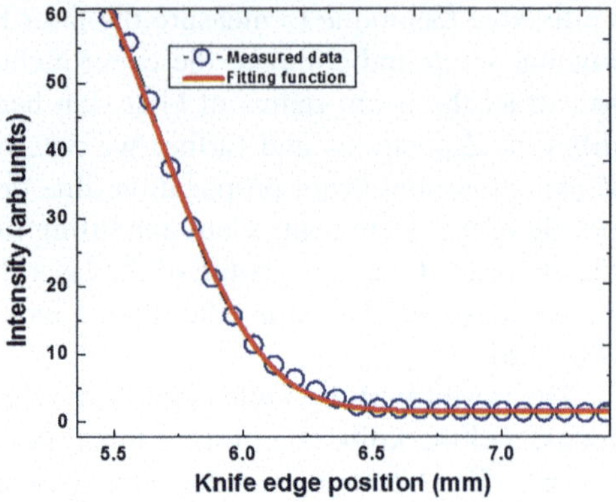

Fig. 2.8 Intensity as a function of position of knife edge at a distance 4.5 cm towards the lens. The open circle symbols are the measured data points while the line is the fitting function obtained by using Eq. 2.1. The estimated value of the beam radius is 621 μm

The beam propagation ratio (M^2) is a value that indicates how close a laser is to a single mode TEM_{00} beam, which in turn determines the beam waist at focus. Figure 2.9 show the M^2 values of −4.318 obtained for the lenses with the focal length of 20 cm, respectively. For the perfect Gaussian TEM_{00} condition, the value of M^2 should be equal to 1. The Gaussian beam profile is determined by M^2 calculation from the values of radii measured at different positions of the knife-edge.

The significance of the beam profile is that the energy density, the concentration, and the collimation of the light are all affected by it. Also the propagation of the beam through space is significantly affected by the beam profile. Since such a variety exists in laser beam profiles, it is essential to measure the profile in any application if the energy distribution affects the performance of the laser or its intended purpose. A typical experimental setup of knife edge technique is shown in Fig. 2.1 of Chap. 2. The knife-edge technique requires a sharp edge (typically a razor blade), a translation stage with a micrometer, and a power meter. In our discussion we consider a radially symmetric Gaussian laser beam with intensity described by

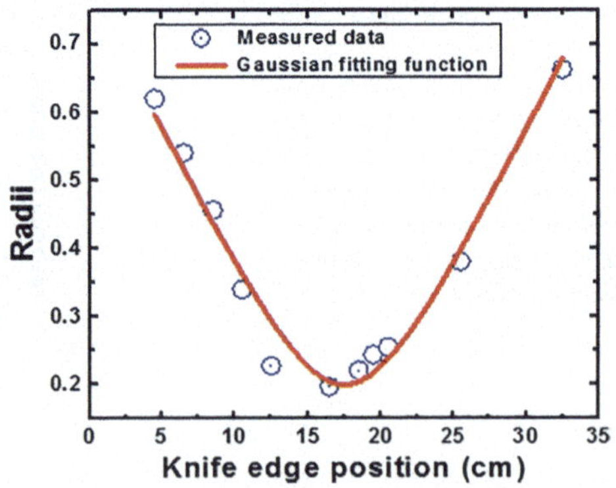

Fig. 2.9 The beam radii as a function of the knife edge position obtained by using plots and various knife edge positions along beam propagation direction (Z-axis) using lens of focal length 20 cm. The value of M^2 is calculated from Gaussian propagation fitting function

$$I(x, y) = I_0 exp[\frac{-(x - x_0)^2 + (y - y_0)^2}{r^2}] \tag{2.3}$$

where I_0 is the peak intensity at the center of the laser beam, located at (x_0, y_0) and x and y are the transverse Cartesian co-ordinates of any point with respect to an origin conveniently chosen at the beginning of the experiment and, w, is the beam radius, measured at a position where the intensity decreases to 1/e times then maximum value of I_0. Equation 2.1 is not the only way to express the intensity of the Gaussian shape laser beam. Some of researchers prefer to define the beam radius at a position where the electric field amplitude drop $1/e$ while the intensity drops to $1/e^2$ times of the maximum value. In 1983 Khosrofian and Garetz [6], described the mathematical procedure by which knife-edge data inverted to yield accurate beam diameter values as well as demonstration of the method using the He–Ne laser. In this technique, with the help of knife-edge initially blocking the laser beam, the laser beam is blocked by the knife edge and the micrometer scale is adjusted to measure the normalized transmittance by the integral,

$$P_N(x) = \frac{1}{2}[1 + erf\frac{(x - x_0)}{r}], \tag{2.4}$$

where erf is the error function. The error function is not an analytical function and it is not used to fit the experimental data during the practical procedure. For the data analysis one of the approach is to take the derivative of Eq. 2.4 which is analytical and is given by

$$\frac{dP_N(x)}{dx} = \frac{1}{\sqrt{\pi r}} exp[\frac{-(x - x_0)^2}{r^2}]. \tag{2.5}$$

But the process of taking derivatives of experimental data with fluctuations results in amplification of the fluctuations and, consequently, an increase in the errors. To overcome this problem, Khosrofian and Garetz suggested a substitution of P_N(x) by an analytical function, which approximately represents P_N(x), to fit the experimental data. The fitting function is given by

$$f(s) = \frac{1}{1 + exp[p(s)]}, \tag{2.6}$$

$$p(s) = \sum_{i=0}^{m} a_i s^i. \tag{2.7}$$

In 2009 Marcos et al., reported the improvement in data analysis by using new polynomial coefficients, however, the method is not used in the present measurements. The laser beam is characterized by using the knife-edge technique and the radius of beam is measured for a convex lens of focal length 10 cm.

In Fig. 2.2 of second chapter, the first measurement of beam radius taken at the focus point lens having (10 cm focal length). The second measurement was taken at

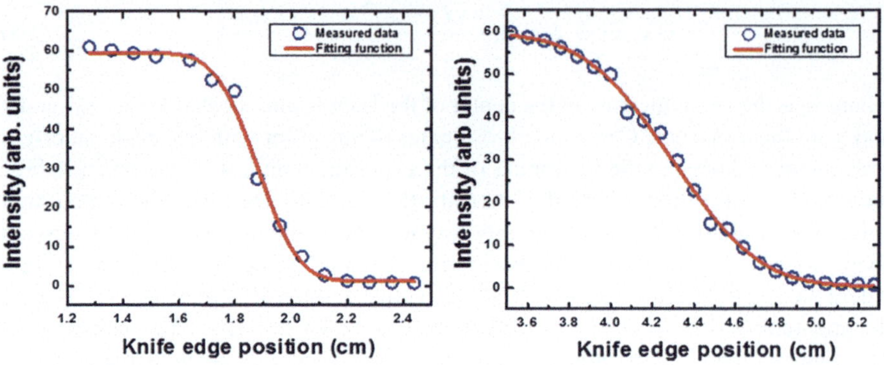

Fig. 2.10 Intensity as a function of position of knife edge at a distance 12.5 cm (left) towards and 32.5 cm (right) away from the lens. The open circle symbols are the measured data points while the line is the fitting function obtained by using Eq. 2.1. The estimated values of the beam radii are 230 and 660 μm for the left and right panels respectively

20 cm from lens/focal point similar process is repeating for beam radius measurement at different position from focal point which is shown in Fig. 2.10.

From Fig. 2.11 one can understand the idea of M^2 profile of Gaussian laser beam. For a perfect Gaussian laser beam, the radii values on the left side R1, R2 and R3 must match the values on the right side with R1', R2' and R3' respectively. The intensity increases with the decrease in the spot size as the intensity of the laser beam is inversely proportional to the radius of beam spot size. The incident laser beam is focused with the help of converging lens. Thereafter, the spot size at different positions along the direction of beam propagation is measured by using knife edge technique. So, the spot size or intensity profile varies at different positions with respect to the focal point. The combination of beam radii at different positions such as in Fig. 2.9 shows the complete profile of laser beam, intensity distribution along z-axis propagate of laser beam, when beam passes through convex lens with fixed focal length. Perfect gaussian beam is required in case of Z-scan to obtain accurate

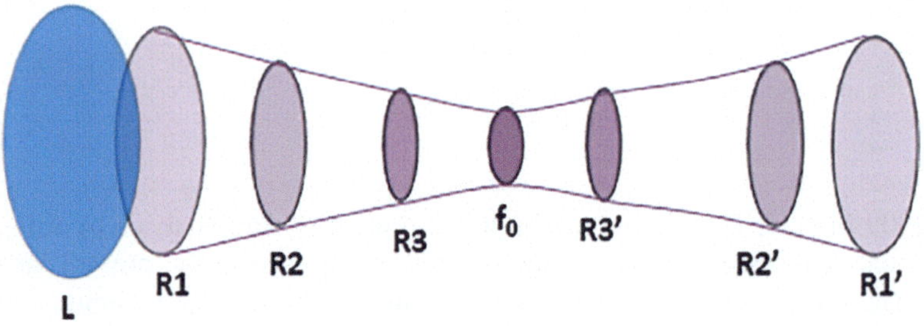

Fig. 2.11 Schematic diagram of beam profile, where L, f_0, R1, R2, R3 and R1', R2', R3' represents the lens, beam radius at focus, radius pre-focus near to the lens and post-focus far from lens

irradiance which is important parameter to extract nonlinear coefficient values of material.

2.2 Temporal Profile Measurements of Ultrafast Pulsed Laser Systems

2.2.1 Introduction to Autocorrelation Technique

Ultrafast laser spectroscopy has become an important tool to investigate material properties for interdisciplinary research where lasers play a dominant role. In solution or gaseous phase, the ultrafast lasers have applications in inspecting and recording transient chemical species which are having a temporary state for very short time interval [11]. The fast reactions are studied under the realm of ultrafast spectroscopy by using short laser pulses.

In mid 1960s, the Nd:glass lasers were discovered and pulsed lasers producing ps pulse-widths were obtained [5]. In 1980, ultrafast laser spectroscopy related research came into picture by the invention of Colliding pulse mode (CPM) color laser, which was used to obtain 100 fs pulses [4, 17] in its initial phase and later on 30 fs pulses were reported [9]. Ultrafast laser pulses are incited by mode-locked lasers and mode locking is a technique in which the laser can be made to produce pulses of light with high intensity and at ps or fs time scales. A short pulse is launched by constructive interference at the point when numerous longitudinal modes are held in phase in a laser resonator. Ti-sapphire is the most commonly used material to produce ultrafast laser pulses [11, 14]. Ti:sapphire can sustain high temperature operation and has a broad emission spectrum range from 700 to 1100 nm centered around 800 nm. The mode locking in nonlinear optical materials is obtained by self-focusing and self-phase modulation. The peak fluence of the ultrafast laser approaches 10^{11} W/cm^2, which is sufficient to observe higher order nonlinearity in the optical medium [?]. Ultrashort laser pulses play a crucial role in nonlinear optics and these lasers can trigger prominent nonlinear phenomena in materials even at a moderate laser power [2].

There are many techniques to measure the temporal profile of the ultrafast pulsed lasers which include autocorrelation, frequency-resolved optical gating (FROG) and spectral phase interferometry for direct electric-field reconstruction (SPIDER). In the case of FROG technique, the autocorrelation traces are obtained as a function of both time delay and frequency, however, the experimental setup of FROG is complex to align. SPIDER is a device that measures the ultrafast pulses utilizing spectral shearing interferometry and can directly give information about the spectral phase.

Autocorrelation is a time resolved technique in which nonlinear crystal is required to produce second-harmonic generation. The autocorrelation technique is useful in the study of the laser spectroscopy, pulse broadening, pulse distortion, fluorescence spectroscopy and Z-scan technique which is used to characterize third and higher order nonlinearity of the materials. In the Z-scan, nonlinear properties are pulse-width dependent, so to measure pulse-width dependent properties of the materials, autocorrelation is an essential tool. We are also discussing the role of nonlinear optical materials in the autocorrelation technique. The ultrafast laser technology includes the use of fs pulsed laser. As an example, the laser system with a pulse-width of 100 fs and the pulse energy of 1 mJ can yield intensity as high as $10^{16} W/cm^2$. At such high power, nonlinear impact becomes progressive. In order to avoid the group velocity mismatch of the incident waves, thin nonlinear crystal are preferred over the thick ones. The aim of the present section is to investigate the temporal profile of ultrafast lasers by using the autocorrelation technique. The various nonlinear optical effects are discussed which is followed by the theoretical background of the autocorrelation technique. The various components of experimental configuration are explored and the nonlinear optical crystal properties are highlighted.

2.2.2 Criteria to Select Nonlinear Optical Materials

In the autocorrelation technique, the nonlinear optical materials play important role, because these materials possess special properties of the phase and the frequency matching conditions [14, 17], when the incident beam is launched at particular angle to the optical axis [2]. The second-harmonic generation is one of the second-order nonlinear optical process in which the incident beam with angular frequency 'ω' is converted into the beam whose angular frequency is '2ω'. This process is spontaneous and involves three photon transitions. It consists of efficient conversion from fundamental to the second-harmonic frequency [12]. However, in order to observe the second-order nonlinear effects, the nonlinear crystal should be non-centrosymmetric in nature.

The coherence of nonlinear optical method is resolved by the nonlinear susceptibility of the medium and the phase-matching parameter. To attain optimal energy conversion efficiency, the phase matching condition $\Delta \mathrm{k} = 0$ has to be fulfilled in many nonlinear optical processes. For the second-harmonic generation, $k = k_1 + k_2$, here k_1 and k_2 are the wave vectors of various frequency components. For frequency doubling, the phase-match parameters, $k^{2\omega} = 2k^{\omega}$. The factor 2 on the right hand side tells about the number of terms contributing to the nonlinear polarization. In SHG, phase matching requires $n(\omega) = n(2\omega)$ [12]. In optics there are different types of autocorrelation methods relying upon what we are estimating [1, 7, 18]. When the electric field is a function of time delay, the autocorrelation function is given below

$$A_{1\tau} = \int E(t) * E(t - \tau)dt \tag{2.8}$$

Here τ is the delay time, and E (t) and E (t-τ) are the electric field amplitudes of original pulse and the optically delayed pulse, respectively. The field autocorrelation is called as first-order autocorrelation. In terms of intensity, the above equation can be written as

$$Ac(\tau) = \int I(t)I(t-\tau)dt \quad (2.9)$$

Here I (t) and I (t-τ) are the intensities of original pulse and the optically delayed pulse, respectively. Equation (5) shows the second-order autocorrelation function, although the higher order correlation function for intensity can also be obtained. To obtain pulse-width from the intensity autocorrelation one must have the knowledge of the pulse shape. For the presumed pulse shape, the autocorrelation width τ_{ac}, the pulse-width τ_p is given by $\tau_{ac} = k\tau_p$. where k is the deconvolution factor. The deconvolution factor is preconceived for systematic pulse shapes or determined numerically for convoluted pulses.

2.2.3 *Experimental Configuration*

An incoming laser beam is converted into two parts by the beam splitter, as shown in Fig. 2.12. In this way, one of the beam has a fixed optical path length while the other beam is directed through an optical delay line (DL). In retro-reflector mirrors, the incoming and reflected beams remain parallel to each other. The first beam is reflected from the spherical mirror and plane mirror, M2, and then directed onto the nonlinear crystal. After the optical delay line, the second beam is reflected by the assembly of mirror M1, spherical mirror and plane mirror M2, and directed onto the nonlinear crystal. It is to be noted that the two beams must have zero optical

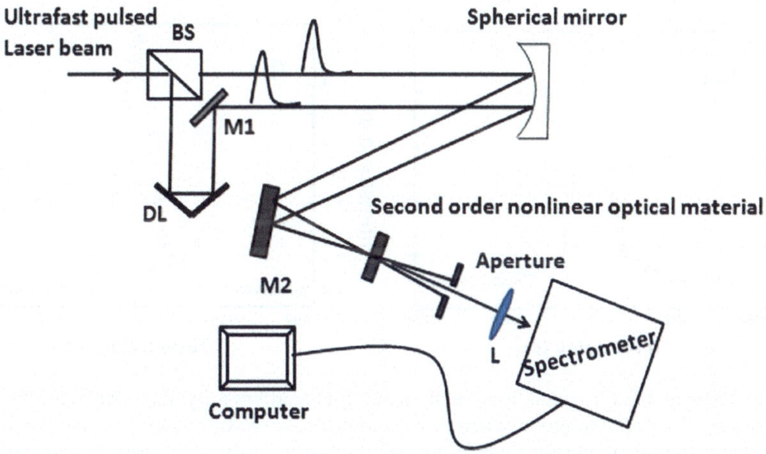

Fig. 2.12 Schematic diagram of experimental configuration of autocorrelation technique. Here BS, DL, M1, M2 and L are beam splitter, optical delay line, mirrors and lenses, respectively

delay when they are incident on the nonlinear crystal. The situation of non-zero optical delay can be optimized by translating the optical delay line. The second-harmonic generation can be observed when the two beams transverse equal optical path lengths. The fundamental beams are also transmitted in this process which are blocked with the help of an aperture. Therefore only the second-harmonic frequency is detected by the detector which is further interfaced to a computer system to record the autocorrelated signal. The delay time can be calculated by $\tau = 2L/c$, where L and c are the optical path length and the speed of light in the vacuum, respectively. The β-barium borate (BBO) and potassium dihydrogen phosphate (KDP) are the nonlinear crystals commonly used to obtain second-harmonic generation, if the phase and frequency matching conditions are satisfied [8, 10]

2.2.4 Estimation of Pulse-Width of Ultrafast Lasers

The second-harmonic generation process is utilized for autocorrelation experiment [15]. We can quantify the full width half maximum (FWHM) of the autocorrelated signal and analyze the width of the power envelope of the pulse by multiplying with 0.71 [13]. The full width at half maximum (FWHM) of the autocorrelation signal is measured with Gaussian fitting function in OriginLab program. The pulse shape decides the value of conversion factor and in Table 3 the values of different conversion factors are given. We use nonlinear curve fitting to fit the curves and examine its pulse shape. Figure 2.13a depicts the autocorrelation trace of a 288 fs pulse obtained by fitting the experimental data with the help of Gaussian fitting function in Origin lab software. A beam pass filter with additional optics is used to record the temporal

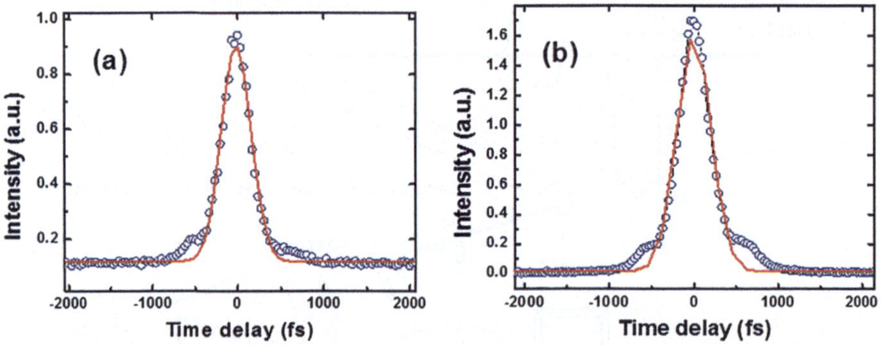

Fig. 2.13 Extracted pulse widths from numerically averaged intensity autocorrelation. The solid line represents the gaussian fitting function used to estimate the full width half maxima and the open circles represents the experimental data. We used origin software to find the pulse width. **a** The FWHM value of 407 fs is obtained by using the fitting function which is multiplied by 0.7071 to deduced the pulse width of 288 fs. **b** The FWHM value of 474 fs is obtained by using the fitting function which multiplied by 0.7071 to deduce the pulse width of 335 fs

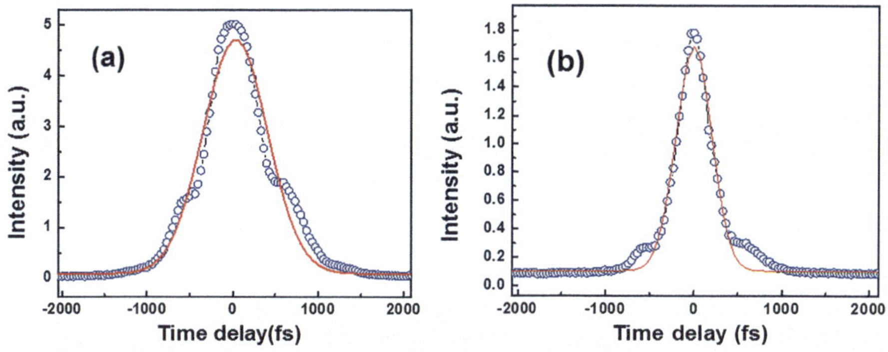

Fig. 2.14 Extracted pulse widths from numerically averaged intensity autocorrelation. The solid line represents the gaussian fitting function used to estimate the full width half maxima and the open circles represents the experimental data. We used origin software to find the pulse width. **a** The FWHM value of 891 fs is obtained by using the fitting function which is multiplied by 0.7071 to deduced the pulse width of 630 fs. **b** The FWHM value of 477 fs is obtained by using the fitting function which multiplied by 0.7071 to deduce the pulse width of 337 fs

Table 2.1 Second-order autocorrelation function and time-bandwidth products for Gaussian pulse-shape model. Here τ_p, τ_G and $\Delta\upsilon$ are pulse width, autocorrelation width and spectral width, respectively

I(t)(x = t/T)	$\Delta\upsilon\,\Delta\,\tau_p$	$\tau_{p/T}$	$G_0^2(\tau)(y=\tau/T)$	τ_G/T	τ_p/τ_G

profile of the ultrafast lasers. However, the sub-panel (b) of Fig. 2.13 illustrates the temporal profile when there is no band pass filter used in optical path. The pulse width of the laser is estimated to be 335 fs. The temporal profile of the ultrafast laser with and without band pass filter in the autocorrelation configuration is shown by Fig. 2.14(a) and (b), respectively. Figure 2.14a, b shows the autocorrelation traces of a 630 fs and 337 fs pulses, respectively. Further, as one can see in Figs. 2.13b and 2.14b, the autocorrelation function presents large wings compare to Fig. 2.13a and 2.14a. These wings may be arises from the chirp pulse, when the laser pulse travel through the different optical components which affect the pulse measurement (Table 2.1).

We have compared different techniques to measure temporal profile of the ultrafast pulsed laser system. The various components of the autocorrelation configuration are elaborated at length. The temporal profile of the ultrafast pulsed lasers are investigated for the cases with and without band pass filter in the experimental setup. The measured pulse-width is found to vary for different configuration, which may be due to the different optical components in the path. It is to be noted that the autocorrelation technique yields estimation of the pulse width, which is equivalent to the pulse width estimation results obtained by other techniques.

We find out the pulse width for the different gaussian pulses using the Origin lab program fitting. We obtain the information of the full-width-half maximum of the ultrashort pulse i.e., the pulse width τ_p. The pulse width τ_p has different values for different shape of the laser pulses.

2.3 Summary

The nonlinear optical experiments depend on the spatial as well as temporal resolution of the utilized laser beam. In the present chapter, the spatial and temoral profile of the ultrashort pulsed laser are investigated by using the knife-edge and the autocorrelation techniques, respectively. Also the experimental techniques to determine the nonlinear optical susceptibility value of the materials are discussed. To calculate the nonlinearity of particular material one must know the linear optical properties of the material such as linear absorption coefficient and linear index of refraction because the one photon absorption (1PA) energy levels can hinder the progress of the nonlinear absorption. The linear optical losses are avoided by working at wavelenth of different from 1PA band to get the off-resonant nonlinearity. The 1PA spectrum of material is determined with the help of UV-Visible spectroscopy and it also cofirms the purity of the sample. In present chapter the temporal and spatial profile measurements of the ultrafast pulsed laser system are discussed. The various components of the autocorrelation configuration are elaborated. The temporal profile of the ultrafast pulsed lasers are investigated for the cases with and without band pass filter in the experimental setup. The measured pulse-width is found to vary for different configuration, which may be due to the different optical components. The various components of the autocorrelation configuration are elaborated at length. The values of laser beam spot size at different positions from the focus are measured with the help of comparison between experimental data and fitting function which is based on the least square fitting.

References

1. A.R. Bakhtari, H.K. Sezer, O.E. Canyurt, O. Eren, M. Shah, S. Marimuthu, A review on laser beam shaping application in laser-powder bed fusion. Adv. Eng. Mater. **26**(14), 2302013 (2024)
2. R.W. Boyd, The nature of the nonlinear optical susceptibility, in *Laser Sources and Applications* (CRC Press, 2020), pp. 1–13
3. M.A. de Araújo, R. Silva, E. De Lima, D.P. Pereira, P.C. De Oliveira, Measurement of gaussian laser beam radius using the knife-edge technique: improvement on data analysis. Appl. Opt. **48**(2), 393–396 (2009)
4. R.L. Fork, B.I. Greene, C.V. Shank, Generation of optical pulses shorter than 0.1 psec by colliding pulse mode locking. Appl. Phys. Lett. **38**(9), 671–672 (1981)
5. U. Keller, K.J. Weingarten, F.X. Kartner, D. Kopf, B. Braun, I.D. Jung, R. Fluck, C. Honninger, N. Matuschek, J.A. Der Au, Semiconductor saturable absorber mirrors (sesam's) for femtosecond to nanosecond pulse generation in solid-state lasers. IEEE J. Sel. Top. Quantum Electron. **2**(3), 435–453 (1996)

6. J.M. Khosrofian, B.A. Garetz, Measurement of a gaussian laser beam diameter through the direct inversion of knife-edge data. Appl. Opt. **22**(21), 3406–3410 (1983)
7. B. Liu, D. Postnov, D.A. Boas, X. Cheng, Dynamic light scattering and laser speckle contrast imaging of the brain: theory of the spatial and temporal statistics of speckle pattern evolution. Biomed. Opt. Exp. **15**(2), 579–593 (2024)
8. V.A. Lobastov, J.D. Ewbank, L. Schäfer, A.A. Ischenko, Instrumentation for time-resolved electron diffraction spanning the time domain from microseconds to picoseconds. Rev. Sci. Instrum. **69**(7), 2633–2643 (1998)
9. P.F. Moulton, Spectroscopic and laser characteristics of TI: Al 2 o 3. JOSA B **3**(1), 125–133 (1986)
10. G. Mourou, S. Williamson, Picosecond electron diffraction. Appl. Phys. Lett. **41**(1), 44–45 (1982)
11. George-Na3240 Porter. Flash photolysis and spectroscopy. a new method for the study of free radical reactions, in *Proceedings of the Royal Society of London. Series A. Mathematical and Physical Sciences*, vol. 200(1061), pp. 284–300 (1950)
12. A. Srinivasa Rao, Overview on second and third order optical nonlinear processes (2016). arXiv:1612.09399
13. D.T. Reid, W. Sibbett, J.M. Dudley, L.P. Barry, B. Thomsen, J.D. Harvey, Commercial semiconductor devices for two photon absorption autocorrelation of ultrashort light pulses. Appl. opt. **37**(34), 8142–8144 (1998)
14. A.E. Siegman, *Lasers* (University science books, 1986)
15. M. Singh, M. Kaur, R. Sharma, Temporal profile measurements of ultrafast pulsed laser systems. Mater. Today: Proc. **26**, 3337–3340 (2020)
16. Y. Suzaki, A. Tachibana, Measurement of the μm sized radius of gaussian laser beam using the scanning knife-edge. Appl. Opt. **14**(12), 2809–2810 (1975)
17. J. Valdmanis, R. Fork, Design considerations for a femtosecond pulse laser balancing self phase modulation, group velocity dispersion, saturable absorption, and saturable gain. IEEE J. Quantum Electron. **22**(1), 112–118 (1986)
18. A. Zarei, S. Pilla, Laser ultrasonics for nondestructive testing of composite materials and structures: a review. Ultrasonics **136**, 107163 (2024)

Chapter 3
Advances in Temporal Profile Measurement of Ultrafast Pulsed Lasers

3.1 FROG

3.1.1 Basic Principle of FROG

FROG is a powerful technique widely used in ultrafast laser spectroscopy and the characterization of ultrashort laser pulses [4]. It can be used to investigate the complex temporal and spectral properties of ultrafast laser pulses, enabling researchers to optimize and control laser sources for various applications. Here, the basic principle of FROG will be discussed to explore its significance in the field of optics and laser science.

FROG is a nonlinear optical method that provides a two-dimensional spectrogram representation of an ultrashort laser pulse which is based on the second-harmonic generation (SHG) autocorrelation combined with a spectral interferometry concept [2]. The FROG experiment begins with the laser pulse to be characterized. The laser pulse is typically split into two identical replicas; referred to as the signal and the reference pulses [3, 5, 8, 10]. The signal pulse undergoes second harmonic generation process to produce a frequency-doubled pulse through a nonlinear crystal. On the other hand, the reference pulse is sent directly to the spectrogram measurement apparatus without any nonlinear interaction. The frequency-doubled signal pulse and the reference pulse then interfere in the spectrogram measurement apparatus. The interference generates a two-dimensional spectrogram that encodes both the spectral and temporal properties of the original ultrashort pulse [6]. This spectrogram is recorded and detected using various techniques such as a charge-coupled device (CCD) camera, a spectrometer, or other time-resolved detection methods. The retrieved spectrogram contains essential information about the spectral intensity as a function of two time variables: the delay time between the signal and reference pulses and the spectral frequency of the signal pulse. By analyzing this spectrogram, researchers can accurately reconstruct the temporal profile and spectral content of the original ultrashort

R. Sharma, *Nonlinear Optical Materials for All-Optical Switching Applications*,
Progress in Optical Science and Photonics 33,
https://doi.org/10.1007/978-981-96-5921-0_3

laser pulse. To extract the temporal and spectral information from the retrieved spectrogram, several numerical algorithms are employed. These algorithms perform an inverse transformation on the spectrogram, converting it back into the temporal and spectral domains of the original pulse. Some popular algorithms include the iterative Fourier transform algorithm, the modified Gerchberg-Saxton algorithm, and the generalized projections algorithm.

The significance of FROG lies in its ability to provide a comprehensive characterization of ultrashort laser pulses, which is essential for understanding and optimizing ultrafast laser sources. Researchers in various fields, including ultrafast spectroscopy, laser physics, and material science, rely on FROG to achieve accurate pulse measurements, allowing them to fine-tune laser parameters for specific applications. Moreover, FROG has applications in exploring and controlling nonlinear optical processes, such as harmonic generation, multiphoton ionization, and coherent anti-Stokes Raman scattering (CARS). These applications are vital in the study of chemical dynamics, biological imaging, and quantum optics, among others [11]. By providing a detailed spectrogram of ultrashort laser pulses, FROG allows researchers to precisely characterize and manipulate laser sources for numerous applications. As technology continues to advance, FROG will undoubtedly remain an indispensable tool in the quest for understanding and harnessing the full potential of ultrashort laser pulses.

3.1.2 Experimental Configuration and Data Analysis of FROG

In the previous section, we have explored the fundamental principles of FROG. Here, we will discuss the experimental configuration and data analysis methods used to retrieve valuable information about the temporal and spectral characteristics of ultrashort laser pulses which are illustrated by Fig. 3.1. The experimental setup

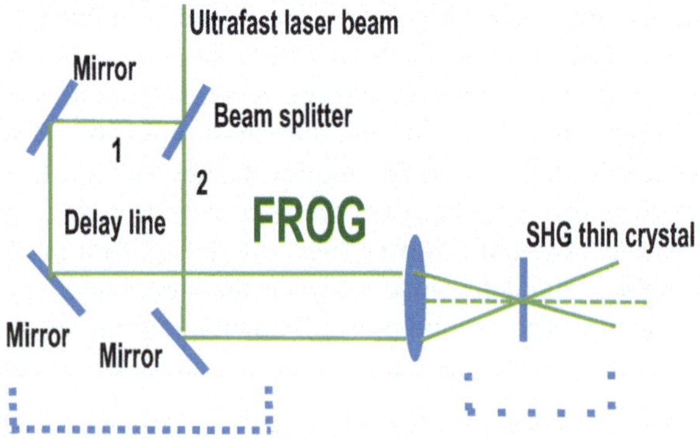

Fig. 3.1 Experimental configuration of the FROG where incoming ultrafast laser beam is split into two parts with the help of a beam splitter. Original incident pulse and the optically delayed pulse are directed onto a thin crystal where the second harmonic generation takes place

of FROG involves several key components that facilitate the generation, manipulation, and measurement of the ultrashort laser pulses. While variations in specific implementations exist, the basic components remain consistent across most FROG configurations. The essential elements of an FROG experimental setup are as follows:

A nonlinear crystal is a crucial component in the FROG setup. The ultrashort laser pulse to be characterized is sent through the nonlinear crystal, where it undergoes a nonlinear optical process, usually second-harmonic generation (SHG). SHG produces a frequency-doubled pulse, which is essential for the generation of the FROG spectrogram. The original ultrashort laser pulse is split into two replicas—the signal pulse and the reference pulse—by using a beam splitter. The signal pulse is sent through the nonlinear crystal, while the reference pulse bypasses the crystal, remaining unchanged. The frequency-doubled signal pulse and the reference pulse are recombined in the spectrogram measurement apparatus, creating an interference pattern that contains the temporal and spectral information of the ultrashort pulse. The apparatus can consist of a spectrometer, a monochromator, or other suitable devices for spectral analysis. The generated interference pattern, or spectrogram, is detected using various devices such as a charge-coupled device (CCD) camera or a photodiode array. The detector converts the optical signal into an electronic signal, which is then recorded for further analysis.

Once the FROG spectrogram is acquired, the data analysis process begins to retrieve the temporal and spectral properties of the original ultrashort laser pulse. Several numerical algorithms are employed for this purpose. Let's explore some of the key steps involved in the data analysis of FROG. Before proceeding with the data analysis, it is essential to calibrate the spectrogram to accurately map the optical delay time and the spectral frequency axes. This calibration is usually performed using a known reference pulse with well-characterized temporal and spectral properties. In the FROG measurement, the recorded spectrogram often contains a background signal, which can arise from various sources such as scattering or stray light. This background signal is subtracted from the recorded data to obtain a clean spectrogram. Numerical algorithms play a vital role in retrieving the pulse information from the FROG spectrogram. Some widely used retrieval algorithms include the iterative Fourier transform algorithm (IFROG), the frequency-resolved optical gating using maximum likelihood (FROG-MAX) algorithm, and the generalized projections algorithm (GP).

a. IFROG: IFROG iteratively modifies the pulse's spectral phase and amplitude until the calculated and measured spectrograms match. It reconstructs the pulse intensity and phase using a series of Fourier transformations.
b. FROG-MAX: FROG-MAX employs a maximum likelihood approach to find the pulse's spectral and temporal characteristics. This algorithm aims to maximize the likelihood that the measured spectrogram corresponds to the reconstructed one.
c. GP: GP is another popular algorithm that finds the pulse's temporal and spectral properties by projecting the measured spectrogram onto a set of basis functions

representing different pulse shapes. The FROG measurement provides information about the pulse's intensity and the spectral phase. To fully characterize the pulse, the spectral phase needs to be retrieved, and this can be performed using algorithms like the Pixel-Encoded Method (PEM).

FROG is an indispensable technique for characterizing ultrashort laser pulses, offering valuable insights into their temporal and spectral properties. The experimental configuration of FROG involves several key components, including ultrafast laser sources, nonlinear crystals, and spectrogram measurement apparatus. After data acquisition, the FROG spectrogram undergoes rigorous data analysis. Calibration, background subtraction, and numerical algorithms, such as IFROG, FROG-MAX, and GP, are employed to retrieve the temporal and spectral characteristics of the original ultrashort laser pulse accurately. With its precise pulse measurement capabilities, FROG continues to be a cornerstone in ultrafast laser spectroscopy and remains essential for advancing research in various fields, including material science, quantum optics, and biological imaging.

3.2 SPIDER

SPIDER is a technique widely used for characterizing ultrashort laser pulses which can provide both amplitude and phase information by exploiting the principles of spectral interferometry.

3.2.1 Basic Principle of SPIDER

The basic principle of SPIDER involves the creation of two identical replicas of the ultrashort pulse to be measured. One of the pulse is optically delayed with respect to the other pulse by using a retroreflector mirror mounted on a delay stage. A chirped pulse is used as a reference which is mixed with the temporally delayed pulse in a nonlinear crystal to obtain sum-frequency generation (SFG) [12]. Overlap of the chirped reference pulse and the optically delayed pulse produces a spectral interference pattern. It is based on the concept of spectral shearing interferometry where the chirped pulse used in SPIDER introduces a frequency-dependent phase shift, or shear, between the two replicas of the incident pulse. This phase shift converts the time-domain information of the pulse into the frequency domain. The resultant spectral interference pattern encodes the phase difference between the two replicas. The interference pattern is then analyzed to extract the spectral phase of the original pulse. This extraction is based on Fourier-transform techniques to convert the measured spectral data into a temporal phase difference. By knowing the fixed time delay between the two pulses, the phase difference can be directly related to the spectral phase of the pulse. One of the key advantages of SPIDER is its ability to directly retrieve the electric field of the pulse without requiring iterative algorithms, which are often necessary in the case of the FROG technique. This direct retrieval makes SPIDER a robust and reliable technique for ultrashort pulse characterization. In practical implementation, a beam splitter is used to create the two replicas of the pulse. These pulse replicas are directed through different optical paths to introduce the desired temporal delay. The chirped reference pulse is generated by passing a portion of the pulse through a dispersive medium to introduce the necessary chirp. The nonlinear interaction for sum-frequency generation typically occurs in a beta barium borate (BBO) crystal.

The output of the nonlinear crystal is analyzed using a spectrometer, which records the spectral interference pattern. The data from the spectrometer is then processed using a computer to extract the phase information and reconstruct the electric field of the pulse. Consequently, SPIDER is a powerful and efficient technique for characterizing ultrashort laser pulses. The method's reliability and precision make it a preferred choice in various applications in ultrafast optics and femtosecond spectroscopy. The direct measurement of temporal and spectral profile of the incident pulse using SPIDER ensures that it remains a vital tool in advancing the understanding and application of ultrashort laser pulses.

3.2.2 Experimental Configuration and Data Analysis of SPIDER

SPIDER is an advanced technique employed to characterize ultrafast laser pulses. This method provides detailed information about the pulse's electric field, including both amplitude and phase. The experimental configuration of SPIDER and its application in ultrafast laser pulse characterization involve several key components and steps [1]. First, a beam splitter is used to split the incident ultrafast laser beam into two parts that need to be characterized. These replicas are subsequently directed through different optical paths, introducing a known optical delay between them. The next component is the chirped reference pulse which is generated by a portion of the original laser pulse sent through a dispersive medium, such as a glass block or a grating pair. This medium introduces a frequency-dependent phase shift, or chirp, to the pulse. The chirped pulse then interacts with the two delayed replicas in a nonlinear crystal, typically beta barium borate (BBO). This interaction results in sum-frequency generation (SFG), where the replicas are nonlinearly mixed with the chirped reference pulse. This nonlinear mixing produces a spectral interference pattern, which is recorded using a spectrometer. This interference pattern is essential for the next step of the SPIDER technique. The spectrometer captures the spectral fringes resulting from the interference of the two temporally delayed pulses. The spectral interference data are processed to extract the phase information where Fourier-transform techniques is used to convert the measured spectral data into temporal phase information. The optical time delay between the two pulse replicas allows for a direct relationship between the phase difference and the spectral phase of the original pulse. The phase information is then used to reconstruct the electric field of the ultrafast pulse.

SPIDER is particularly valuable in fields that require precise characterization of ultrafast laser pulses. In ultrafast optics, for instance, understanding the exact electric field of a pulse is crucial for experiments involving nonlinear optics and time-resolved spectroscopy. The ability to accurately characterize the amplitude and phase of ultrashort pulses enables researchers to design and optimize experiments with greater precision. In femtosecond spectroscopy, SPIDER helps in studying ultrafast processes by providing detailed pulse information. This capability is essential for probing rapid dynamic events in materials science, chemistry, and biology. The detailed characterization of laser pulses ensures that the temporal resolution of the spectroscopy experiments is maximized.

Overall, the experimental configuration of SPIDER, involves the creation of optically delayed pulse replicas, the generation of a chirped reference pulse, and the nonlinear mixing in a crystal which allows characterization of ultrafast laser pulses with precision. The technique's ability to directly retrieve the electric field without iterative methods makes it a robust and reliable tool in various applications requiring detailed pulse characterization. By providing accurate phase and amplitude information, SPIDER significantly advances the study and application of ultrafast laser pulses in multiple scientific disciplines.

3.3 GRENOUILLE

Recalling the fact that the autocorrelation technique to measure the temporal profile of the ultrafast laser pulses involves mainly two aspects; the first point is that a single pulse is split into the two pulses with the help of a beam splitter which are to be overlapped in time, the so called the zero delay. The second point is to take care of the spatial overlap of the ultrafast pulses during the entire measurement. Moreover, a thin second-harmonic generation crystal is used which also results into weak intensity of the resulting beam. All of these points make the autocorrelation experimental configuration difficult to align. Therefore, there is a strong requirement of an experimental configuration which can give results of complete pulses with minimized optical alignment efforts. In the forthcoming subsection, basic principle, experimental configuration and data analysis of GRENOUILLE technique will be highlighted.

3.3.1 Basic Principle of GRENOUILLE

GRENOUILLE is a simplified version of FROG experimental configuration which can used to obtain complete information of the ultrafast pulses. It is to be noted that the optical alignment of various elements is relatively difficult in the case of FROG and SPIDER, however, minimum efforts are required to align the GRENOUILLE [9].

3.3.2 *Experimental Configuration and Data Analysis of GRENOUILLE*

Experimental configuration of the FROG is illustrated in Fig. 3.2a where incoming ultrafast laser beam is split into two parts with the help of a beam splitter. Laser beam 1 passes through the optical delay line which is further directed onto a thin crystal for obtaining the second harmonic generation. The laser beam 2 is reflected from a mirror and focused onto the thin crystal through a lense. The beam 1 and 2 are overlapped in space and time in the thin crystal which produces the second harmonic generation signal [7]. GRENOUILLE experimental configuration is depicted in Fig. 3.2b where the optical elements of FROG on the left such as beam splitter, optical delay line and mirrors etc. are replaced by a single element; a Fresnel biprism, while the thin crystal of FROG is replaced by a thick crystal in GRENOUILLE.

Recalling the fact that the Fresnel biprism is typically used to obtain the division of the wavefront of the incoming beam so that interference takes place, however, GRENOUILLE does not deal with the interference aspects. On the top portion of the Fresnel biprism, the first pulse comes before the second pulse while both of the pulses arrives at the same time in the central portion. In the bottom part of the Fresnel

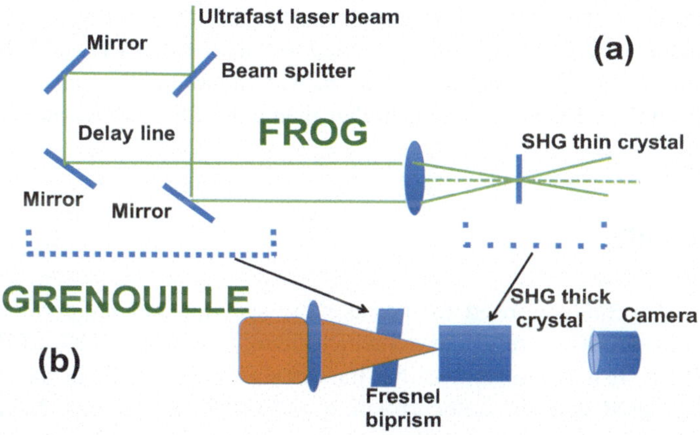

Fig. 3.2 **a** Experimental configuration of the FROG where incoming ultrafast laser beam is split into two parts with the help of a beam splitter. Original incident pulse and the optically delayed pulse are directed onto a thin crystal where the second harmonic generation takes place. **b** GRENOUILLE experimental configuration where the optical elements of FROG on the left are replaced by a Fresnel biprism while the thin crystal of FROG is replaced by a thick crystal in GRENOUILLE. Taken from the study of swam optics https://www.swampoptics.com/grenouille.html

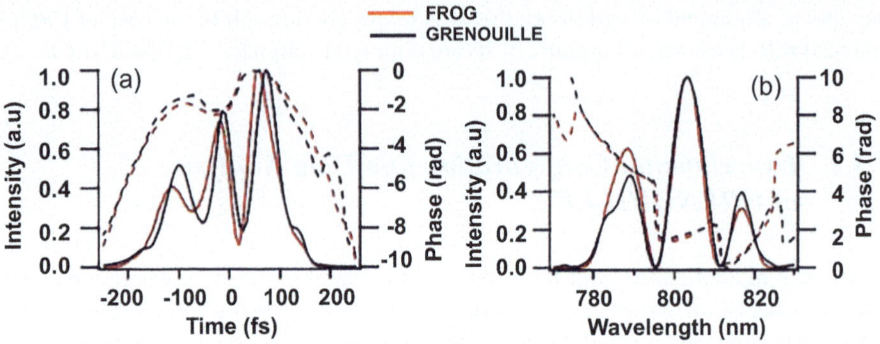

Fig. 3.3 **a** Temporal and **b** spectral profile of a complex pulse measured by GRENOUILLE and FROG techniques. Taken from the study of swam optics https://www.swampoptics.com/grenouille.html

biprism, the first pulse comes after the second pulse. Consequently, the broad range of the optical delays can be obtained for a single pulse without any tedious optical alignments. There is a difference between the spectrum obtained from the thin and thick crystals. The thin crystal gives broad second harmonic spectrum while thick crystal separates the colors which can be used as a spectrometer.

It is quite evident that the ultrafast pulses without any chirp can be easily measured with the autocorrelation and FROG techniques. On the other hand, it is relatively difficult to estimate the temporal and spatial profiles of the ultrafast pulses if there is chirping in the pulses especially the double chirped ones. The test results of the temporal profile of the double chirped pulse are highlighted in Fig. 3.3a where the precision and accuracy of the results obtained by GRENOUILLE and FROG technique are comparable. The spectral domain measurements are illustrated by Fig. 3.3b the signal from the GRENOUILLE is higher than that of the FROG measurements.

3.4 Summary

The advanced techniques used to measure the spectral and temporal profile of the ultrafast laser pulses are discussed which mainly include FROG, SPIDER and GRENOUILLE. It is demonstrated that the ultrafast pulses without any chirp can be easily measured with the autocorrelation and FROG techniques. Nevertheless, it is relatively difficult to estimate the temporal and spatial profiles of the ultrafast pulses if there is chirping in the pulses especially the double chirped ones. For such cases, GRENOUILLE technique produces best results with minimum number of the optical elements and alignment problems.

References

1. M.E. Anderson, A. Monmayrant, S.-P. Gorza, P. Wasylczyk, I.A. Walmsley, Spider: a decade of measuring ultrashort pulses. Laser Phys. Lett. **5**(4), 259 (2008)
2. D. Daiya, R.K. Patidar, A. Moorti, M.P. Kamath, N.S. Benerji, All reflective frog using near 90° retro-reflecting mirror for single shot intensity and phase profile measurement of ultra-short laser pulse. Res. Opt. **15**, 100661 (2024)
3. A. Dulat, A.D. Lad, C. Aparajit, A. Choudhary, Y.M. Ved, L. Veisz, G. Ravindra Kumar. Single-shot, spatio-temporal analysis of relativistic plasma optics. Optica **11**(8), 1077–1084 (2024)
4. X. Gu, Z. Wang, S. Akturk, R. Trebino, Using frog to measure spatio-temporal distortions in ultrafast laser beams. In: *CLEO/Europe. 2005 Conference on Lasers and Electro-Optics Europe*, pp. 363– 2005 (2005)
5. P. Huang, H. Yuan, H. Cao, H. Wang, X. Wang, Y. Wang, W. Zhao, F. Yuxi, Direct sampling of ultrashort laser pulses using third-harmonic generation. Opt. Lett. **48**(8), 2154–2157 (2023)
6. S.I. Hwang, C.H. Nam, W. Cho, Temporal characterization of femtosecond laser pulses using terahertz time-domain spectroscopy. Sci. Rep. **9**(1), 15234 (2019)
7. D. Lee, P. Bowlan, A.P. Shreenath, P. Gabolde, Q. Cao, S. Akturk, X. Gu, R. Trebino, Ultra-simple alignment-free ultrashort pulse measurement devices. In: *Commercial and Biomedical Applications of Ultrafast Lasers VI*, vol. 6108. (SPIE, 2006), pp. 256–262
8. Z.-W. Lu, X.-D. Hou, F. Wan, Y.I. Salamin, C. Lv, B. Zhang, F. Wang, Z.-F. Xu, J.-X. Li, Diagnosis of ultrafast ultraintense laser pulse characteristics by machine-learning-assisted electron spin. Matter Radiat. Extrem. **8**(3) (2023)
9. P. O'Shea, M. Kimmel, G. Xun, R. Trebino, Highly simplified device for ultrashort-pulse measurement. Opt. Lett. **26**(12), 932–934 (2001). (Jun)
10. Tata Institute of Fundamental Research. A 'measure' advance for ultrashort laser light. Optica (2024). https://phys.org/news/2024-08-advance-ultrashort-laser.html
11. R. Trebino, R. Jafari, S.A. Akturk, P. Bowlan, Z. Guang, P. Zhu, E. Escoto, G. Steinmeyer, Highly reliable measurement of ultrashort laser pulses. J. Appl. Phys. **128**(17) (2020)
12. S. Zheng, Y. Cai, X. Pan, X. Zeng, J. Li, Y. Li, T. Zhu, Q. Lin, X. Shixiang, Two-step phase-shifting spider. Sci. Rep. **6**(1), 33837 (2016)

Chapter 4
Advanced Nonlinear Optical Techniques

4.1 Criteria to Select Nonlinear Optical Materials

In the autocorrelation technique, the nonlinear optical materials play important role, because these materials possess special properties of the phase and the frequency matching conditions [18, 19], when the incident beam is launched at particular angle to the optical axis [4]. The second-harmonic generation is one of the second-order nonlinear optical process in which the incident beam with angular frequency 'ω' is converted into the beam whose angular frequency is '2ω'. This process is spontaneous and involves three photon transitions. It consists of efficient conversion from fundamental to the second-harmonic frequency [15]. However, in order to observe the second-order nonlinear effects, the nonlinear crystal should be non-centrosymmetric in nature. The coherence of nonlinear optical method is resolved by the nonlinear susceptibility of the medium and the phase-matching parameter. To attain optimal energy conversion efficiency, the phase matching condition $\Delta k = 0$ has to be fulfilled in many nonlinear optical processes. For the second-harmonic generation, $k = k_1 + k_2$, here k_1 and k_2 are the wave vectors of various frequency components. For frequency doubling, the phase-match parameters, $k^{2\omega} = 2k^{\omega}$. The factor 2 on the right hand side tells about the number of terms contributing to the nonlinear polarization. In SHG, phase matching requires $n(\omega) = n(2\omega)$ [15]. In optics there are different types of autocorrelation methods relying upon what we are estimating. When the electric field is a function of time delay, the autocorrelation function is given below

$$A_{1\tau} = \int E(t) * E(t - \tau)dt \tag{4.1}$$

Here τ is the delay time, and E (t) and E (t-τ) are the electric field amplitudes of original pulse and the optically delayed pulse, respectively. The field autocorrelation is called as first-order autocorrelation. In terms of intensity, the above equation can be written as

$$Ac(\tau) = \int I(t)I(t - \tau)dt \tag{4.2}$$

R. Sharma, *Nonlinear Optical Materials for All-Optical Switching Applications*, Progress in Optical Science and Photonics 33,
https://doi.org/10.1007/978-981-96-5921-0_4

Here I (t) and I (t-τ) are the intensities of original pulse and the optically delayed pulse, respectively. The Eq. (4.5) shows the second- order autocorrelation function, although the higher order correlation function for intensity can also be obtained. To obtain pulse-width from the intensity autocorrelation one must have the knowledge of the pulse shape. For the presumed pulse shape, the autocorrelation width τ_{ac}, the pulse-width τ_p is given by $\tau_{ac} = k\tau_p$. Where k is the deconvolution factor. The deconvolution factor is preconceived for systematic pulse shapes or determined numerically for convoluted pulses.

4.2 Nonlinear Characterization Technique

4.2.1 Single Arm Z-scan Experiment

In this section, we start our discussion with measurement of the $\chi^{(3)}$ nonlinearity in materials by using single-arm Z-scan technique. The DA Z-scan is discussed to address the issue of nonlinear optical measurements of the solute molecules if the solvent has significant contribution to the nonlinear signal. The section ends with mathematical description of the FOM values to decide selection of an appropriate material for the AOS applications. The higher order nonlinear effects come into play when the amplitude of the optical field applied across the material is increased. There are many techniques to investigate NLR value of the materials which include nonlinear interferometry [12, 20], degenerate three and four-wave mixing [1, 8], ellipse rotation [14], and beam distortion measurement [16, 21]. Here we discuss Z-scan technique to analyze the third-order nonlinear susceptibility of the materials which was developed by Bahae et al. in 1985 [17]. It is a single beam method with spatial beam distortion which has the sensitivity of around $\lambda/500$. The Z-scan technique can be simultaneously used to measure the NLA and the NLR of the materials. In the Z-scan technique, a single gaussian beam with tight focus geometry is used to obtain the NLR and the NLA coefficient values. In the case of the NLR measurements, the transmittance of nonlinear medium through a finite aperture in the far field is measured as a function of the sample position with respect to the focal plane. The aperture is not required in the case of the NLA measurement although rest of the setup is similar to that of the NLR measurements. The inset of Fig. 4.1a shows the general block diagram of the Z-scan measurement where the incident Gaussian beam enters from left which is further splitted into two parts with the help of a 90:10 beam splitter (not shown). About 10% of the original signal is directed onto a reference detector to reduce noise in the measured signal. Refer to the main panel of Fig. 4.1a, the signal on detector without any aperture corresponds to the NLA in the material and is referred as an open aperture (OA) scan measurements. An aperture is placed in the far field and the transmitted intensity through the specimen is measured with the help of a detector (Fig. 4.1b). The aperture is closed to 30% of the total intensity transmitted to the detector. Refer to the inset of Fig. 4.1b the sample is translated along the Z-position and the signal is referred to as a closed aperture

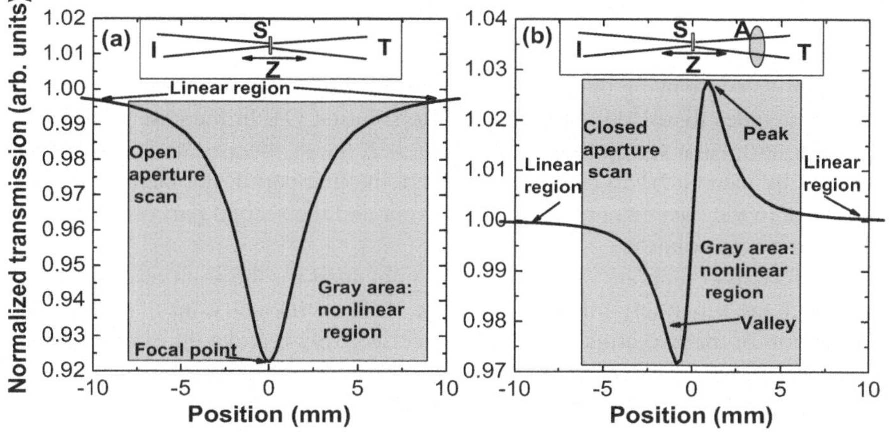

Fig. 4.1 The sensitive single-arm Z-scan traces for **a** the open aperture (OA) and **b** the closed aperture (CA) signals. The inset. (**a**) is a general example of the experimental configuration to record the OA signal. In the inset, the incident Gaussian beam (I) is focused onto the thin sample (S) which translates along the Z-direction (Z) while the transmitted signal (T) is recorded by using a detector. (**b**) The CA signal representing the Z-scan curve with a valley-peak configuration which is a signature of the positive value of the nonlinear refractive index coefficient. Here inset. of (**b**) is a general example of the experimental configuration to record the CA signal which is similar to the OA configuration except for the fact that an aperture (A) is inserted before the signal is detected by the detector. The transmitted signal is flat in the linear region while the nonlinear region is included in the gray box. Note the optical nonlinearity increases as the sample translates to the focal point where the beam irradiance is maximum. The details are referred to the text and Ref. [17]

(CA) signal in our forthcoming discussion. The situation is highlighted in the main panel of Fig. 4.1b where the curve with a pre-focal valley and a post-focal peak is a signature of positive value of the NLR. It is to be mentioned that the NLR signal has some contribution of the NLA signal, therefore, both OA and CA methods are used to correct the measured values by dividing CA signal by the OA signal. In the present case, the imaginary nonlinear susceptibility of the material corresponds to the NLA coefficient while the real part gives information about the NLR coefficient values. The NLR coefficient value is measured by the phase shift of the incident beam at the focus. In this way, the normalized transmittance, S, is fixed to 0 and 1 for the CA and the OA signals, respectively. The situation can be utilized to measure the NLA coefficient value (α_2) where main contribution to NLA signal is from the two-photon absorption. Therefore, Z-scan technique can be considered as the single-beam method with no tedious alignment which yields the magnitude as well as the sign of the χ^3 via separation of the real and the imaginary parts. The entire principle of the Z-scan utilizes the wavefront distortion due to the nonlinearity of the sample, therefore, scattering from the sample and surface inhomogeneties can significantly affect the Z-scan signatures. Consequently, the samples having good surface quality are required to perform experiments with this technique.

4.2.2 Dual Arm Z-scan Experiment

The researchers are using the dual-arm Z-scan to investigate the NLR values of the solute molecules dissolved in an appropriate solution [7]. In the case of dual-arm Z-scan experimental set-up illustrated by Fig. 4.2, the incident beam is divided into two parts by using a 50:50 beam splitter. Here the first part of the incident beam is directed onto the cuvette containing the solvent and the second part of the beam is directed onto the solution.

Therefore, two separate Z-scan traces are obtained for the solution and the solvent which are separately filled to obtain the NLR coefficient values. Note that the equalization of the two arms is necessary to effectively remove the contribution of the solvent signal. Initially the pulse-width, energy and beam waist are equalized in the two arms which is followed by the equalization of Z-position of the samples and the aperture linear transmittance.

The third-order nonlinear optical susceptibility is the sum of the real and the imaginary parts and is given by following equation

$$\chi^3 = Re(\chi^{(3)}) + Im(\chi^{(3)}) \tag{4.3}$$

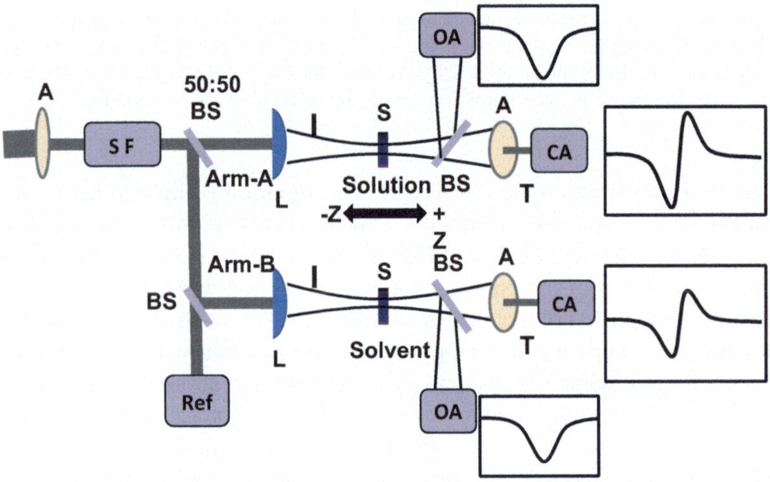

Fig. 4.2 Schematic illustration of the DA Z-scan experimental configuration. The various components are the lenses (L), the apertures (A), the spatial filter (SF), the beam splitters (BS), the sample (S), the reference detector (Ref), the open-apertures (OA), and the closed-apertures (CA) [6]

The real part of the nonlinear optical susceptibility can be related to the NLR values by relation;

$$Re(\chi^{(3)}) = 2{n_0}^2\epsilon_0 c n_2 \quad (4.4)$$

The imaginary part corresponds to the NLA,

$$Im(\chi^{(3)}) = \frac{{n_0}^2\epsilon_0 c^2\alpha_2}{\omega} \quad (4.5)$$

The NLR and the NLA coefficients of the materials can be measured by above listed equations which can be further used to decide the figure of merit (FOM) of the NLO material. The FOM is useful quantity for the AOS applications and can be written as;

$$FOM = |\frac{Re(\chi^{(3)})}{Im(\chi^{(3)})}| \quad (4.6)$$

$$FOM = \frac{4\pi n_2}{\lambda_0\alpha_2} \quad (4.7)$$

where, λ_0 is the wavelength of the incident beam. But we can not measure the transient NLR and NLA effects in materials by using single and dual-arm Z-scan. Hence, the beam deflection technique was used to determine n_2 values with the $\lambda/20,000$ accuracy [2, 6, 22].

In order to fabricate the high figure-of-merit materials, one of methods was highlighted by using selenopyrylium end group to increase the conjugation length of the molecule design of the pi-conjugated organic materials without any change of the symmetry of the bond length. This type of the design of the pi-conjugated organic materials along with the selection of the photon energy to reduce the linear and nonlinear losses can significantly enhance the FOM values. The details of the fabrication process are referred to the Refs. [9, 10]. On the other hand, the chemical precipitation method was used to synthesize the NQDs where the CdO, the hexadecylamine, the Se powder, the trioctylphosphine oxide, the trioctylphosphine, and the stearic acid were used to prepare the precursor solutions. The surface of the CdSe was passivated by alkylamine ligands to obtain high quantum yield. The precursors of CdO and stearic acid were heated in a flask to 200 °C and cooled down to room temperature. Thereafter, the trioctylphosphine oxide and the hexadecylamine were added and the solution was again heated. The Se solution and the trioctylphosphine were mixed together, and injected to the solution prepared in the previous step [11].

4.2.3 Pump-Probe Experiment

The pump-probe measurements are carried out to investigate the effects of the pump photon energy and the pulse-width on the material. This technique is used to determine transient absorption response of the carrier at different wavelengths of light. The ultrafast carrier dynamics of the material is measured with weak probe light because of the photon energy of the probe beam is below the interband transition threshold of material. When the pump pulse impinging on the specimen, the free electrons inside go to excited state, leading to an increased transmission of probe beam. The transmission reaches its peak point at the end of the pump beam. The material under investigation gradually relax to the original state which is captured by the probe pulse with reduced transmission.

The charge carrier dynamics of the materials were probed by the ultrafast laser spectroscopy. Figure 4.3 demonstrates the experimental configuration for ultrafast femtosecond transient absorption spectroscopy, a Ti-Sapphire femtosecond laser system at 800 nm wavelength. The pulse width, repetition rate and the energy per pulse values of the fs laser system are 35 fs, 1 KHz and 4 mJ respectively. The beam is directed onto an optical parametric amplifier used to produce required excitation wavelengths of pump beam. The remaining 0.5 mJ of energy is directed to delay line of 0–8 ns which is further directed to the spectrometer. The beam is further made incident onto the sapphire crystal for white light generation so that the probe beam wavelength can be selected [5]. The output beam can be tuned in a wavelength range of 240–2600 nm with the help of OPA system. One of the beam will act as the pump beam.

4.2.4 Nonlinear Absorption

Normally, the photons are absorbed in metals and semiconductor materials. Metals absorb most of frequencies because valance and conduction bands overlap, however, there is an energy gap in the case of semiconductors. The phenomenon is entirely different in semiconductor materials because selective frequencies are absorbed. In

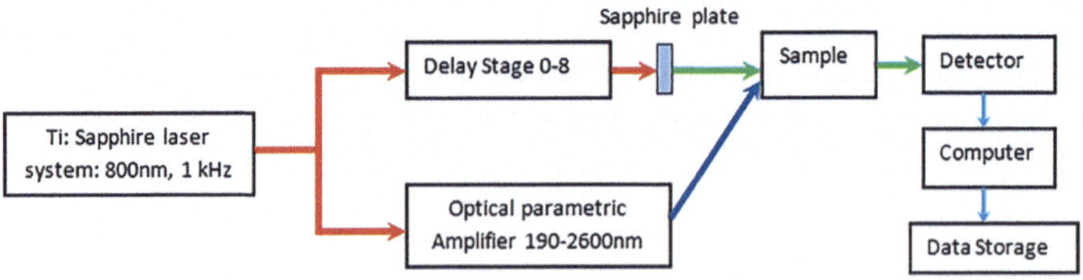

Fig. 4.3 Schematic diagram of ultrafast transient absorption spectroscopy

the case of semiconductor the incident photon is absorbed when it has sufficient energy to excite an electron from the valence band to the conduction band, otherwise, incident frequency will be transmitted. In linear absorption, only a single photon is responsible for electronic excitation. Nevertheless, more than one photon is absorbed in two-photon and multiple photon absorption. It means two or more photon absorption is required to excite the electron from valence band to conduction band. The question arises how two-photon are absorbed. And why we need to study this nonlinear absorption and its applicability.

4.3 Nonlinear Refraction and Absorption Measurement of GaAs, CS2 and Fused Silica Materials

4.3.1 *Effects of Nonlinear Absorption*

The present technique is based on the principle of spatial beam distortion. In this technique, the laser beam will be distorted at every point in the space over z-position. The Gaussian beam divided into small slices in the space along z-axis where the beam intensity at every point is different.

Thus, once an open aperture (S = 1) Z-scan is performed [3, 13], the nonlinear absorption coefficient can be obtained [16]. An experimental procedure shown in Fig. 4.1 of Chap. 2, where 0.5 mm thick GaAs sample is examined using 185 fs (FWHM) pulses at 1550 nm wavelength. GaAs with band gap energy of 1.42 eV is two-photon absorber at this wavelength. With linear refractive index is 3.3702, the diffraction length inside the sample ($n_0 z_0$) was approximately twice than the sample thickness. This allows us to safely apply the thin sample approximation. Figure 4.4 depicts the open aperture scan data at peak irradiance I_0 of 0.451 GW/cm^2 and also theoretical result plotted in Fig. 4.5. Here we simulate the experiment by using computational tool MATLAB, which is used to determine the nonlinear absorption coefficient by inserting input parameters for the program. To obtain nonlinear absorption coefficient values we need simulations, however, the aperture is removed to allow all the light to reach the detector. This sets the normalization transmittance to S = 1. This is used to measure the nonlinear absorption coefficient, α_2. The main cause of NLA is due to two-photon absorption or multi-photon absorption processes. The refractive index, linear transmittance, etc. are the same, but prediction of NLA is different. All these results are extracted from simulation by comparing the experimental data shown by blue circular symbols with theoretical curve in shown in red color. The chi-square fitting along with the slowly varying envelope approximation is used to obtain nonlinear absorption coefficient [17]. For instance, Fig. 4.4 represents the data obtained from experiment; the nonlinear absorption coefficient value can not be predicted only by experimental data. One needs to take all the parameters of the experimental set-up like; pulse energy of laser, irradiance at every position (z), radius of curvature of Gaussian beam, radial and temporal components of beam.

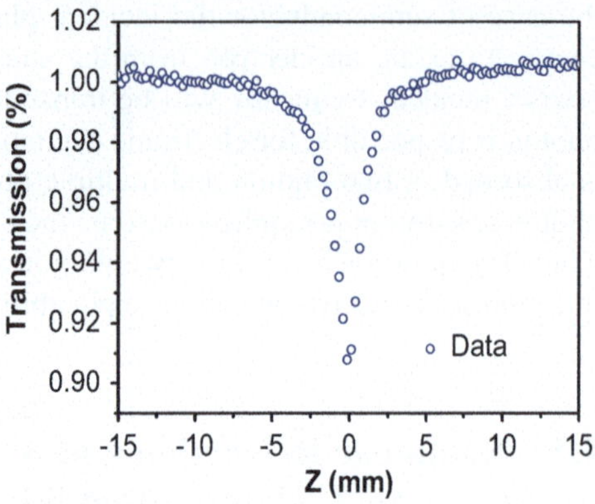

Fig. 4.4 The experimental data of z-scan transmittance of 0.5 mm thick GaAs sample measured using femtosecond laser pulses at $\lambda = 1550$ nm with $I_0 = 0.451$ GW/cm^2 pulse energy E = 1 nJ. The curve is obtained without any aperture

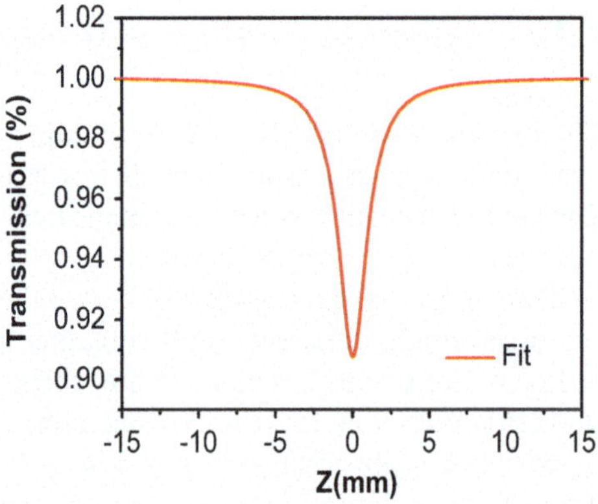

Fig. 4.5 The theoretical fit for z-scan transmittance of 0.5 mm thick GaAs sample measured using femtosecond pulses at $\lambda = 1550$ nm with $I_0 = 0.451$ GW/cm^2 pulse energy E = 1 nJ. The solid line is the theoretical result. The curve is obtained without any aperture and predicted value of α_2 is 13.5 cm/GW

All the data is inserted into MATLAB code basis [17] to obtain the value of nonlinear absorption coefficient. The theoretical curve will be changed when nonlinear absorption coefficient is changed, but experimental absorption curve will be fixed. At last the theoretical curve is fitted to the experimental curve to obtain the nonlinear absorption coefficient value of GaAs sample under investigation. Figures 4.6, 4.7, 4.8 and 4.9 shows the nonlinear absortion values obtained at different pulse energy and beam irradiance values.

The low-energy background (LEB) is not taken into account because the signal to noise ratio in GaAs is very high. But in solutions, one would have low-energy background. LEB is a noise signal present in materials due to the linear optical phenomenon or diffraction of incident beam.

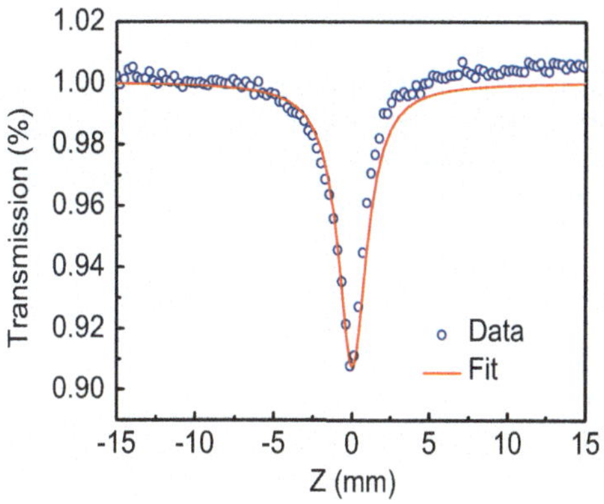

Fig. 4.6 Normalized z-scan transmittance of 0.5 mm thick GaAs sample measured using femtosecond pulses at $\lambda = 1550$ nm with $I_0 = 0.451$ GW/cm^2 pulse energy E = 1 nJ. The solid line is the theoretical fit. The curve is obtained without any aperture and predicted value of α_2 is 13.5 cm/GW

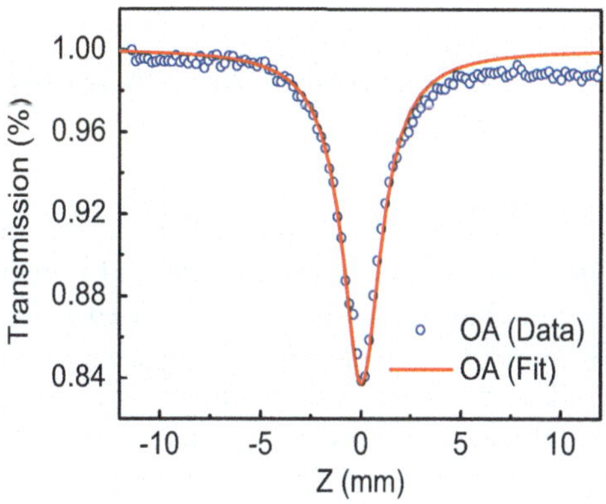

Fig. 4.7 Normalized z-scan transmittance of 0.5 mm thick GaAs sample measured using femtosecond pulses at $\lambda = 1550$ nm with $I_0 = 0.676$ GW/cm^2 pulse energy E = 1.5 nJ. The curve is obtained without any aperture and predicted value of $\alpha_2 = 18$ cm/GW. Blue dots shows the experimental data taken from z-scan and red curve represent the theoretical fit

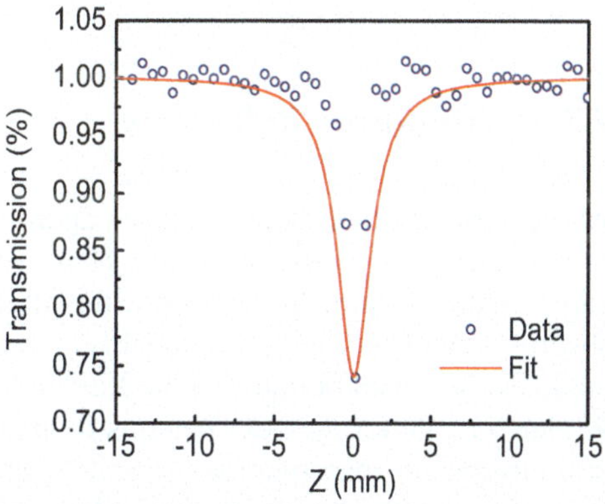

Fig. 4.8 Normalized z-scan transmittance of 0.5 mm thick GaAs sample measured using femtosecond pulses at $\lambda = 1550$ nm with $I_0 = 0.832$ GW/cm^2 pulse energy E = 1.846 nJ. The curve is obtained without any aperture and predicted value of $\alpha_2 = 28$ cm/GW. Blue dots shows the experimental data taken from z-scan and red curve represent the theoretical fit

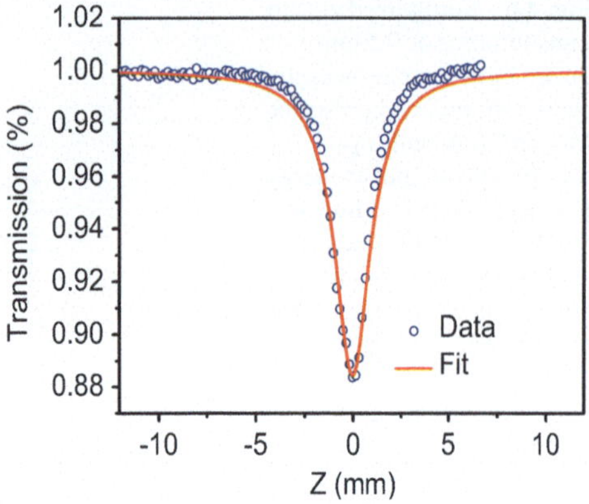

Fig. 4.9 Normalized z-scan transmittance of 0.5 mm thick GaAs sample measured using femtosecond pulses at $\lambda = 1550$ nm with $I_0 = 0.361$ GW/cm^2 pulse energy E = 0.8 nJ. The curve is obtained without any aperture and predicted value of $\alpha_2 =$ 22 cm/GW. Blue dots shows the experimental data taken from z-scan and red curve represent the theoretical fit

4.3.2 Discussion of Results on the Basis of Laser Pulse Energy

All the results are taken at different laser pulse energy in Figs. 4.6, 4.7, 4.8 and 4.9 and these results shows that the pulse energy affect the nonlinearity of materials. The nonlinear absorption is increased with increase in the pulse energy of the laser. Because higher the incident energy higher will be the perturbation of molecules of the nonlinear crystals. Semiconductors possess various nonlinear optical mechanisms, mainly due to the presence of free-carrier state and ease with which free carriers may be photo-generated. In degenerate four wave mixing the output signal is generated either via coherence at harmonics of the input waves or via a nonlinearity due to intensity dependent changes in the absorption coefficient in the material. In Figs. 4.6, 4.7, 4.8 and 4.9 results contain absorption coefficient at different pulse energies. All experimental data (blue dots) are taken at different energy [7, 16].

4.3.3 Nonlinear Refraction

We performed experiment of closed aperture Z-Scan theoretically using MATLAB programming to get the n_2 value. Here we have used CS_2 as sample with optical path lengths of 1 mm. To determine the intensity dependent nonlinear refraction and absorption with Z-scan technique this is an essential requirement to keep the sample thickness less than or equal to the Rayleigh range. In the present investigations, the Rayleigh range for the laser beam is found to be 1 mm. The CS_2 sample is inserted in a cuvette for investigations. The CS_2 sample is exposed to an intense beam of femtosecond Laser having wavelength of 1550 nm to produce the nonlinearity in CS_2. We have varied energy of the incident beam in order to see the effect of the incident beam intensity on the nonlinear refractive index. We obtained these results through

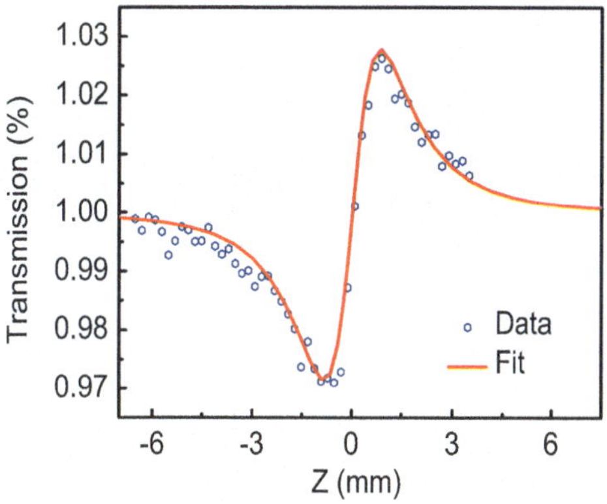

Fig. 4.10 Measurement of nonlinear refraction in one arm of Z-scan setup of CS_2 at 50 nJ pulse energy

numerical simulations of the n_2 values whose experimental data was obtained from College of Optics and Photonics, (CREOL) USA. The curves shown in Fig. 4.10 are the result of Arm A of a dual arm closed aperture Z-Scan experiment in which CS_2 sample of thickness 1mm is exposed to the a femtosecond laser beam of energy 50 nJ and wavelength 1550 nm. As shown in Fig. 4.11 we have an experimental data (blue dot) for nonlinear measurements fitted to the theoretical data (red curve) to obtain the nonlinear refractive index coefficient n_2 value 1.8×10^{-15} cm^2/GW. The values of incident energy and wavelengths are 50 nJ and 1550 nm, respectively. The curves shown by Fig. 4.12 are the result of Arm B of a dual arm closed aperture Z-Scan experiment in which CS_2 sample of thickness 1 mm is exposed to the a femtosecond laser beam. In the dual arm Z-scan, the peak-to-valley transmission has to be matched in both arms. The curves shown by Fig. 4.12 are the result of Arm A of a dual arm closed aperture Z-Scan experiment For Arm A nonlinear refraction coefficient (n_2) value of 3.4 e^{-15} cm^2/GW is obtained. We took all the parameters of the experimental set-up into cosideration like; pulse energy of laser,

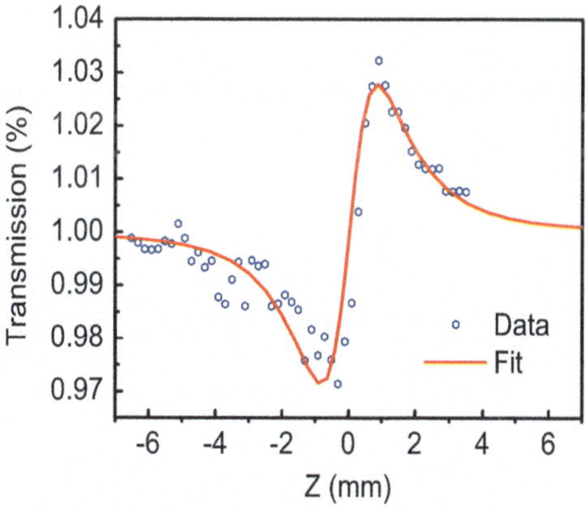

Fig. 4.11 Measurement of nonlinear refraction in other arm of Z-scan setup of CS_2 at 50 nJ pulse energy

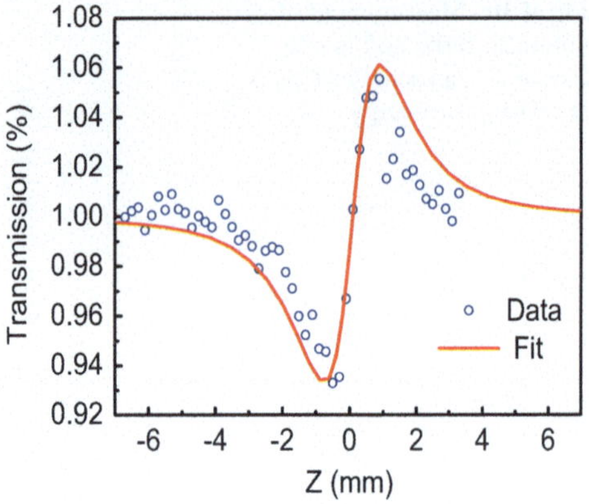

Fig. 4.12 Measurement of nonlinear refraction in one arm of Z-scan setup of CS_2 at 60 nJ pulse energy

irradiance at every point of position (z), radius of curvature of beam, radial and temporal components of beam. All this information was inserted in the MATLAB program which helped to calculate complex mathematical equation in minimum time. We altered the values of nonlinear refraction coefficient and at a particular point, the theoretical curve can be fitted with the experimental curve.

All optical switch is a device that allows to control an optical signal with other optical signal. In vacuum two light beams just pass through each other without interacting. So in vacuum it is not possible to change the direction of a light beam with the other beam. On the other hand in a nonlinear optical material intense incident light beam changes the optical properties of the material which affects any other beam passing through the material. Therefore first intense beam can control the second beam during interaction with the material. Figures 4.13, 4.14 and 4.15 represents the NLR signature of 1.6 mm thick fused silica sample which can be used as calibration material.

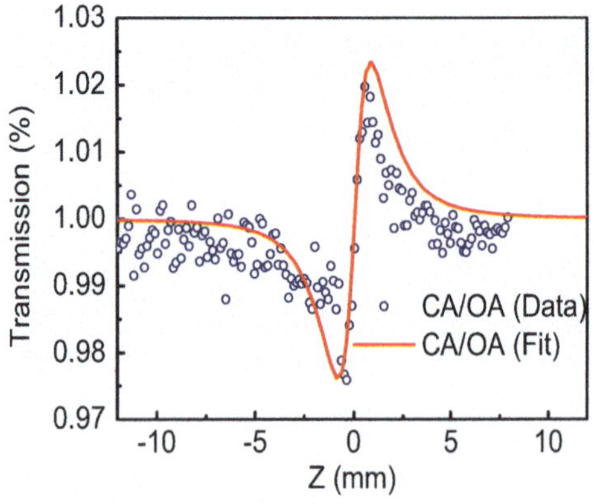

Fig. 4.13 The nonlinear refraction of 0.5 mm thick fused silica sample is measured using femtosecond laser pulses at $\lambda = 1550\,\text{nm}$ with $I_0 = 172.9\ \text{GW/cm}^2$ pulse energy E = 280 nJ. The obtained nonlinear refraction coefficient value is $n_2 = 4.8\times10^{-16}\ \text{cm}^2/\text{GW}$

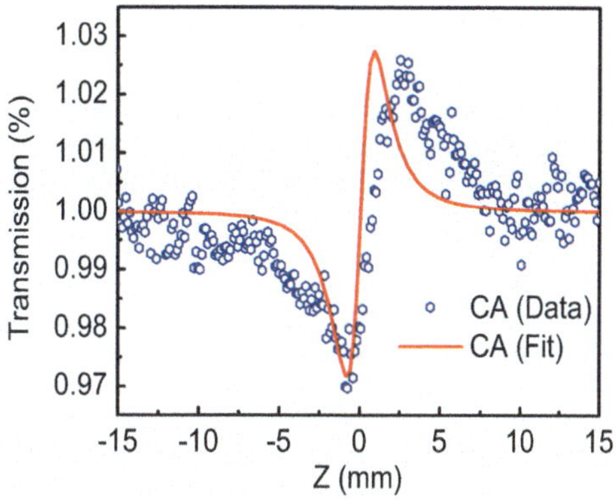

Fig. 4.14 Measurement of nonlinear refraction of 1.6 mm thick fused silica sample using femtosecond pulses at $\lambda = 1550$ nm with $I_0 = 96.330$ GW/cm^2 pulse energy E = 156 nJ. The obtained nonlinear refraction coefficient value is $n_2 = 4.7 \times 10^{-16}$ cm^2/GW

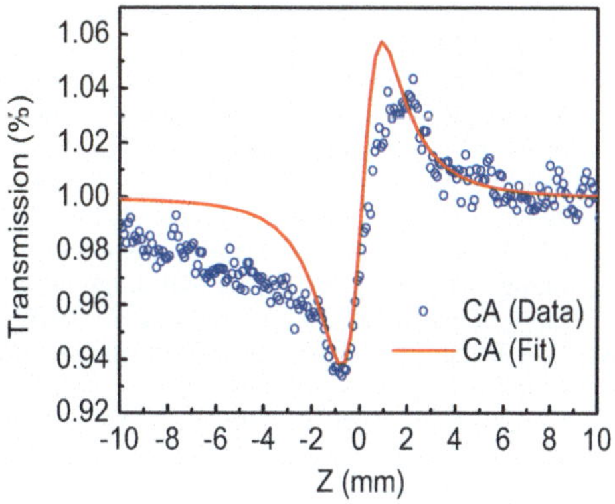

Fig. 4.15 Measurement of nonlinear refraction of 1.6 mm thick fused silica sample using femtosecond pulses at $\lambda = 1550$ nm with $I_0 = 122.265$ GW/cm^2 pulse energy E = 198 nJ. The obtained nonlinear refraction coefficient value is $n_2 = 3$ e^{-16} cm^2/GW

4.4 Summary

Also the experimental techniques to determine the nonlinear optical susceptibility value of the materials are discussed. To calculate the nonlinearity of particular material one must know the linear optical properties of the material such as linear absorption coefficient and linear index of refraction because the one photon absorption (1PA) energy levels can hinder the progress of the nonlinear absorption. The linear optical losses are avoided by working at wavelength of different from 1PA band to get the off-resonant nonlinearity. The 1PA spectrum of material is determined with the help of UV-Visible spectroscopy and it also confirms the purity of the sample. Here we demonstrate simulation based on the theoretical model to measure the nonlinear absorption coefficient of GaAs material. In case of single photon absorption GaAs in infrared region (873 nm), on the other hand in two photon absorption it will be absorb visible region (436 nm). The investigations of the present thesis can be used for applications of optical detectors, two photon absorption and all optical switching etc.

References

1. R. Adair, L.L. Chase, S.A. Payne, Nonlinear refractive-index measurements of glasses using three-wave frequency mixing. JOSA B **4**(6), 875–881 (1987)
2. H.-S. Albrecht, P. Heist, J. Kleinschmidt, D.V. Lap, Ultrafast beam-deflection method and its application for measuring the transient refractive index of materials. Appl. Phys. B **57**, 193–197 (1993)
3. H. Ali, A. Anjum, D. Goswami, Investigating the third-order nonlinear optical properties of a schiff base (DBAP) and its co/cu metal complex using z-scan and dft methodology. J. Mol. Struct. **1307**, 138039 (2024)
4. R.W. Boyd, The nature of the nonlinear optical susceptibility, in *Laser Sources and Applications*. (CRC Press, 2020), pp. 1–13
5. P. Devi, P. Kumar, C. Sharma, M. Kumar, M. Nayak, Structural and ultrafast charge carrier dynamics of in-situ synthesized rGO/Fe3O4 nanocomposite. Mater. Lett. **181**, 99–102 (2016)
6. M.R Ferdinandus, H. Hu, M. Reichert, D.J. Hagan, E.W. Van Stryland, Beam deflection measurement of time and polarization resolved ultrafast nonlinear refraction. Opt. Lett. **38**(18), 3518–3521 (2013)
7. M.R. Ferdinandus, M. Reichert, T.R. Ensley, D.A. Fishman, S. Webster, D.J. Hagan, E.W. Van Stryland, Measuring small solute nonlinearities in solution by dual-arm z-scan technique, in *Frontiers in Optics* (Optica Publishing Group, 2012), pp. FM4I–6
8. STEPHENR Friberg and PETERW Smith, Nonlinear optical glasses for ultrafast optical switches. IEEE J. Quantum Electron. **23**(12), 2089–2094 (1987)
9. J.M. Hales, J. Matichak, S. Barlow, S. Ohira, K. Yesudas, J.-L. Brédas, J.W. Perry, S.R. Marder, Design of polymethine dyes with large third-order optical nonlinearities and loss figures of merit. *Science* **327**(5972), 1485–1488 (2010)
10. S.A. Haque, J. Nelson, Toward organic all-optical switching. Science **327**(5972), 1466–1467 (2010)
11. S. Kim, R. Sharma, B. Kim, H.-S. Yang, K. Kyhm, Modal gain enhancement by cylindrical waveguide and gain saturation in cdse nanocrystal quantum dots. J. Phys. D: Appl. Phys. **42**(9), 095403 (2009)
12. M. Moran, C.-Y. She, R.J.I.J. Carman, Interferometric measurements of the nonlinear refractive-index coefficient relative to cs 2 in laser-system-related materials. IEEE J. Quantum Electron. **11**(6), 259–263 (1975)
13. E. Mzwd, S.K. Alsaee, N. Suardi, A.A. Aziz, S. Arshad, Exploring the nonlinear optical properties of gold nanoparticle colloids capped with gum arabic: A study utilizing continuous wave laser and z-scan technique. Opt. Mater. **148**, 114907 (2024)
14. A. Owyoung, Ellipse rotation studies in laser host materials. IEEE J. Quantum Electron. **9**(11), 1064–1069 (1973)
15. A. Srinivasa Rao, Overview on second and third order optical nonlinear processes (2016). arXiv:1612.09399
16. M. Sheik-Bahae, Sensitive n2 measurements using a single beam, in *Laser-Induced Damage in Optical Materials 1989*, vol. 1438, pp. 148–157 (SPIE, 1990)
17. M. Sheik-Bahae, A.A. Said, T.-H. Wei, D.J. Hagan, E.W. Van Stryland, Sensitive measurement of optical nonlinearities using a single beam. IEEE J. Quantum Electron. **26**(4), 760–769 (1990)
18. A.E. Siegman. *Lasers* (University Science Books, 1986)
19. J. Valdmanis, R. Fork, Design considerations for a femtosecond pulse laser balancing self phase modulation, group velocity dispersion, saturable absorption, and saturable gain. IEEE J. Quantum Electron. **22**(1), 112–118 (1986)
20. M.J. Weber, D. Milam, W.L. Smith, Nonlinear refractive index of glasses and crystals. Opt. Eng. **17**(5), 463–469 (1978)
21. W.E. Williams, M.J. Soileau, E.W. Van Stryland, Optical switching and n2 measurements in cs2. Opt. Commun. **50**(4), 256–260 (1984)
22. P. Zhao, M. Reichert, S. Benis, D.J. Hagan, E.W. Van Stryland, Temporal and polarization dependence of the nonlinear optical response of solvents. Optica **5**(5), 583–594 (2018)

Chapter 5
Optical Nonlinearities in Semiconductors

5.1 Two-Photon Absorption in Semiconductors

The nonlinearities of the semiconductor materials can be exploited for various applications in optical technology and are useful to fabricate important components of the radiation detectors in the linear and nonlinear regimes of optics. The semiconductor nonlinearities were extensively investigated to understand the Kerr effect, NLR with free-carrier generation, 2PA, and free-carrier absorption effects relying on the NLA. On the other hand, the optical limiting and the mid-IR detection in semiconductors were reported by different researchers [4, 11]. There are multiple mechanisms which are responsible for the nonlinearities in the semiconductors, such as anharmonic response of the bound electrons, the third-order optical nonlinearity owing to the bound electrons in the crystalline solid materials. These effects dominate in the pure semiconductors at the optical frequencies below the bandgap energy. The NLO characteristics were modeled by various researchers by using a one-dimensional anharmonic oscillator problem [3]. In the case of moderately doped semiconductors, dominant third-order NLO susceptibility can be obtained by the nonlinear motion and the free-carrier energy relaxation caused by the incident electric field. The nonlinearities are found to be dominant in the narrow-band semiconductors owing to the large non-parabolicity of the conduction band. The free-carrier nonlinearities can be further classified into three types: (1) D-2PA, (2) ND-2PA, and (3) END-2PA.

R. Sharma, *Nonlinear Optical Materials for All-Optical Switching Applications*,
Progress in Optical Science and Photonics 33,
https://doi.org/10.1007/978-981-96-5921-0_5

5.2 Non-Degenerate Two-Photon Absorption in Semiconductors

In the case of D-2PA, two equal energy photons are absorbed when an electron makes a transition from the valance band to the conduction band, which is shown in the left portion of Fig. 5.1 [6]. On the other hand, two photons of different energies excite the electron from the valence band to the conduction band in the ND-2PA.

5.3 Extremely Non-Degenerate Two-Photon Absorption

In the case of END-2PA [8], the energy of one of the incident photons is larger, while the second photon is of extremely low energy. In Fig. 5.2a, the pump-probe measurements were carried out to investigate the effect of the pump photon energy and the pulse width on the ND-2PA in the semiconductor materials. The ND-2PA in the CdTe sample was obtained by using ps pulses with a pump energy of about 9.7% of the energy bandgap. This low pump frequency allows the variation of the probe energy ($h\omega_e$); 910 nm (94%), 900 nm (95%), 890 nm (96%), 880 nm (97%), 870 nm (98%) of the bandgap energy. The pump pulse excites the electrons from the valence band to the intermediate states, which is followed by the probe or signal pulse to excite the carriers from the intermediate state to the conduction band. It is relatively difficult to guess which frequency excites the electron from the ground to the intermediate or the intermediate state to the higher state. In two-photon detectors, the magnitude of the END-2PA remains the same irrespective of the fact that the signal and the gate wavelengths are altered.

The researchers had used low-frequency pump photons to obtain a variety of signal frequencies over a wide range; however, a narrow range is selected by using low-energy transitions and the linear absorption edge [1]. Both the pump and the

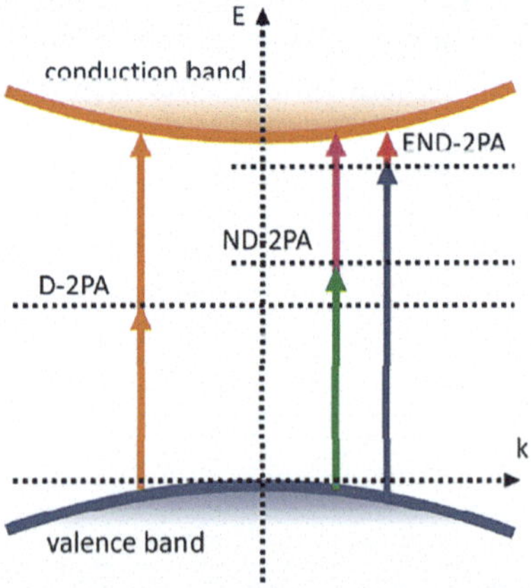

Fig. 5.1 Schematic representation of transitions involved in a 2PA process for the photons having various energy ratios. The ND-2PA and END-2PA are characterized by a small detuning energy as compared to the bandgap [1]

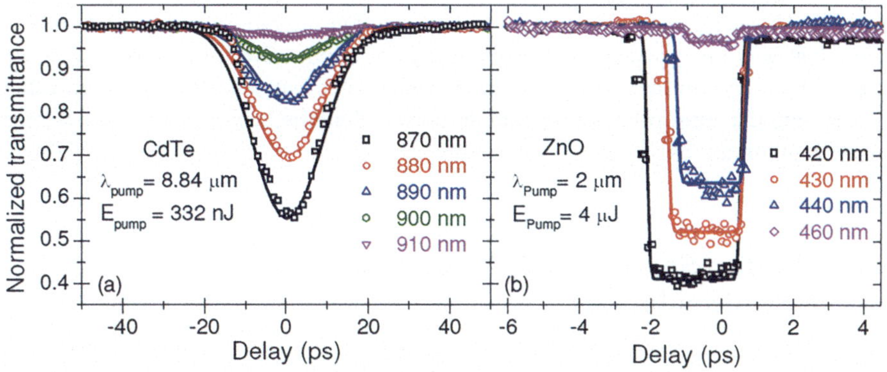

Fig. 5.2 The pump-probe data in **a** the CdTe samples measured by using picosecond pulses, while (b) the ZnO sample is measured by using femtosecond pulses. The solid lines are fitted to the experimental data represented by various symbols [6]

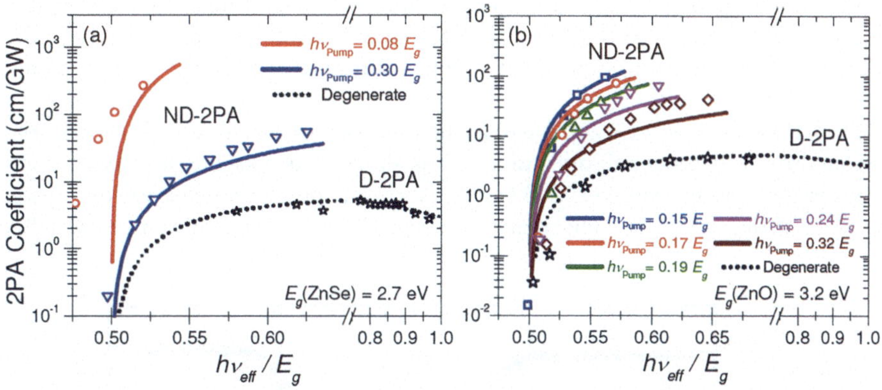

Fig. 5.3 Comparison of nondegenerate 2PA spectra of **a** the ZnSe sample and **b** ZnO recorded with the help of the femtosecond laser pulses. The nondegenerate and degenerate spectra are illustrated by the solid and the dashed lines, respectively [1]

probe frequencies are responsible for the transitions between the energy band gap of the materials under study. The signal photon energy is varied by the difference between the bandgap energy verses the pump energy.

The group velocity dispersion (GVD) played an important role when the ZnO semiconductor was excited by utilizing the fs laser pulses [1]. It is quite evident from the temporal profile of the pump-probe data in Fig. 5.2b. The pump-probe walk-off takes place during the propagation through the sample because, for the initial delays, the group velocity strongly depends on the wavelength. The comparison of 2PA spectra of ZnSe and ZnO for the degenerate and nondegenerate cases using femtosecond laser pulses is illustrated by Fig. 5.3. Note that about a 270 times enhancement of the nondegenerate 2PA was seen in comparison to the degenerate case for ZnSe. The pump-probe energies are selected in such a way that the pump energy is about 8% of the bandgap energy while the probe energy is about 92%. In a similar way, we propose that the nonlinear refractive index can also be tuned to the maximum value by selecting the appropriate pump and probe wavelengths, which can be suitable for the AOS applications.

5.4 Ultrafast Spectroscopic Measurements of Semiconductors

For GaAs semiconductor, the energy band gap is 1.42 eV. The transmission as a function of the optical time delay at 1300 nm wavelength is illustrated in Fig. 5.4, where pump-probe energy is fixed at 20nJ and the pulse width of the incident beam is 150 fs. The probe transmission shows a constant value when the time delay is between −600 fs to around −250 fs with a large number of data points. In the temporal measurements, there must be zero delay between the pump and probe signals. The flat region of the transmission signal signifies that pump and probe signals are not overlapped [5, 7].

From the time delay of −200 fs to 0 fs, the value of the transmission drops to 0. The sudden drop in the transmission value at 0 fs shows that transient absorption dominates in this region. After a −200 fs delay time, the probe transmission decreases, which is attributed to the enhancement of bound-electronic nonlinearity and ESA

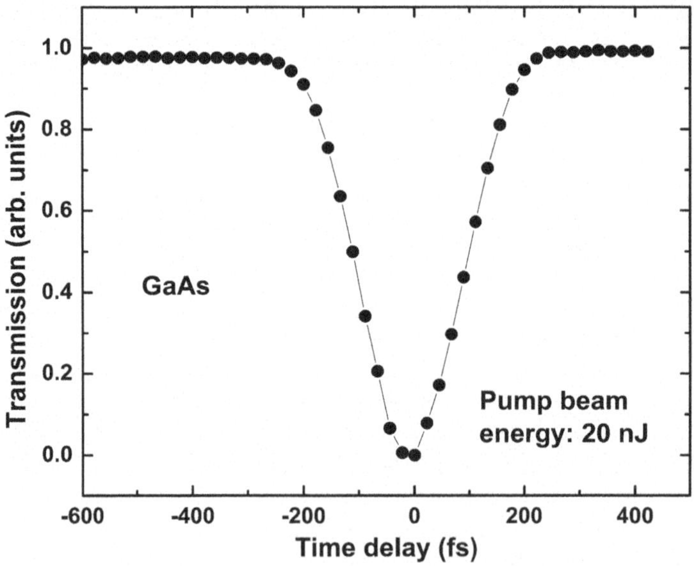

Fig. 5.4 The pump-probe data for the GaAs semiconductor at 20 nJ pump beam energy

femtosecond time regime. The value of transmission is around 0.99 at −400 fs, 0.85 at around −200 fs, and then there is a sudden decrease in the value of transmission. Further, the value of transmission reaches zero at 0 fs. As the time delay passes 0 fs, there is a sudden increase in the transmission value, and the initial value of transmission is achieved within 200 fs to 400 fs because the fast nonlinear absorption region is crossed. It means the probe does not capture or is not affected by the carriers that are excited by the pump beam. The pump-probe data for the GaAs semiconductor with a 20 nJ pump beam energy show a significant change in transmission with time delay. The constant value of transmission in the initial time delay scale and further sudden decrease shows that the ultrafast nonlinear optical phenomenon plays a significant role in GaAs. The knowledge of optical responses is necessary for understanding the ultrafast optical properties of GaAs for their potential applications in ultrafast all-optical signal processing (AOSP).

The pump-probe measurements of the GaAs with 48 nJ pump beam energy are illustrated in Fig. 5.5, where a relatively less number of points are taken to see if the effect is still visible. In this plot, the transmission is represented in arbitrary units with a scale from 0.84 to 1.0, and the time delay is shown in the fs with a scale from −600 to 400. The transmission shows a mixed trend of value when the time delay is between −600 fs and around −250 fs.

The value of the transmission starts to decrease from 0.98 to around 0.94 between the time delay of −250 fs and −180 fs. Further, a decrease in the value of transmission in the range of -600 fs to -200 fs is observed. The value of transmission drops to 0.94 and 0.87 at −180 fs and −90 fs, respectively. The lowest value of transmission is approximately 0.85 around 0 fs time delay. Thus, at 0 fs the transmission is at its

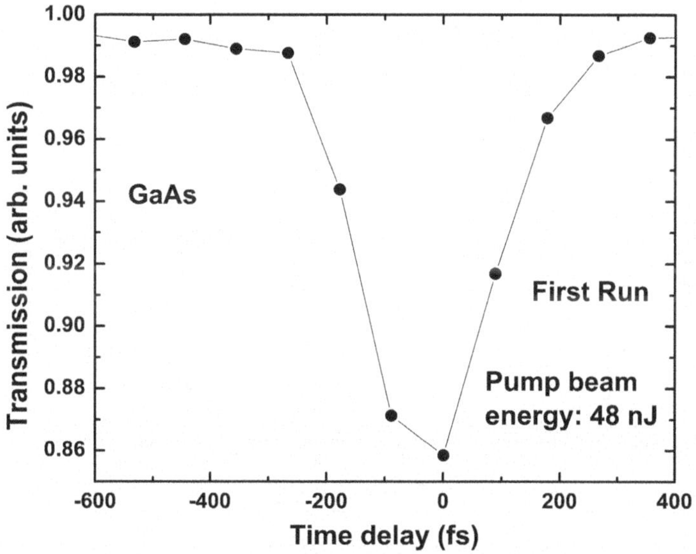

Fig. 5.5 The pump-probe data for the GaAs semiconductor at 48 nJ pump beam energy

lowest, so the absorption phenomenon dominates in this region. Further, the value of transmission shows a sudden increase in the transmission value from 0 fs to 400 fs, and a value similar to the initial transmission is attained within this time delay range. The pump-probe data for the GaAs semiconductor with a 48 nJ pump beam energy shows a significant change in transmission over time delay, showing the highest value of transmission with 0.99 at around 450 fs, and the lowest value of 0.85 at around 0 fs, showing that at this point, absorption dominates transmission effects.

A second run of pump-probe measurements at 48 nJ is carried out, and the details are highlighted in Fig. 5.6 with an increased number of points. In this plot, the transmission is represented in arbitrary units with a scale from 0.80 to 1.00, and the time delay is shown in the fs with a scale from −600 to 600 approximately. The transmission shows a mixed trend of value when the time delay is between −600 fs to around −300 fs. In this range, the value of transmission is high when a pump beam is induced with an energy of 48 nJ. The value of the transmission starts to decrease from 0.99 to around 0.97 between the time delay of −280 fs and −180 fs, respectively. Further, a sudden decrease in the value of transmission in the range of −180 fs to 0 fs. The value of transmission is approximately 0.95 at −155 fs, dropped to 0.91 at −124 fs, and further dropped to 0.84 at −62 fs. Further, 0.81 is the lowest value of transmission obtained at 0 fs. Thus, at 0 fs the transmission is at its lowest, so the absorption phenomenon dominates in this region. Further, the value of transmission shows a sudden increase in the transmission value from 0 fs to 600 fs, and a value similar to the initial transmission is achieved within this time delay range. Two important points can be understood from Figs. 5.5 and 5.6.

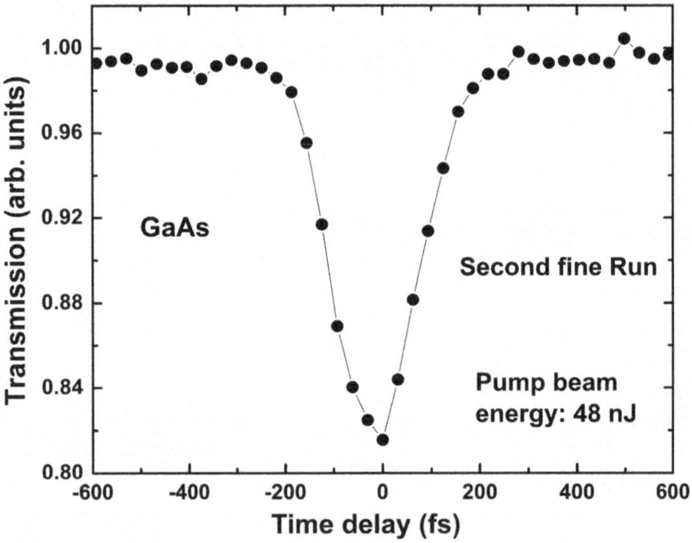

Fig. 5.6 The pump-probe data for the GaAs semiconductor repeated experimental data for more accuracy at 48 nJ pump beam energy

(1) The pump-probe results are reproducible even if the number of points is increased and there are no additional peaks for nonlinear absorption.
(2) The symmetry of the two-photon absorption can be affected, which indicates inhomogeneity of the sample.

Figure 5.7 shows the pump-probe data for the GaAs semiconductor at different pump beam energies represented with 20 nJ, 48 nJ, and 48 nJ_{fine}. The transmission signal remains flat when the time delay is between −600 fs to around −200 fs with a large number of data points in each case. Further, when the time delay is between −200 fs and around 0 fs, the value of transmission is highest in the 48 nJ fine represented with a blue triangle data point curve, and the lowest value is obtained at 20 nJ shown with a black rectangle data point curve. So, GaAs at 20 nJ shows the least transmission as compared to other values of pump beam energy. Therefore, absorption is high in this case, hence the value of 2PA is high at 20 nJ, so it can be utilized in optical limiting applications.

Next, we investigate pump-probe measurements of CdSe, which are illustrated in Fig. 5.8. The energy band gap of CdSe is 1.74 eV, and it has been investigated by many researchers for luminescence-related applications. Nevertheless, it would be interesting to investigate the two-photon absorption through CdSe by using the degenerate wavelength of 1300 nm for pump-probe beams. In this plot, the transmission is represented in arbitrary units with a scale from 0.70 to 1.05, and the time delay is shown in fs with a scale from −600 to 600 fs. The transmission shows a mixed trend of value when the time delay is between −600 fs and around −400 fs.

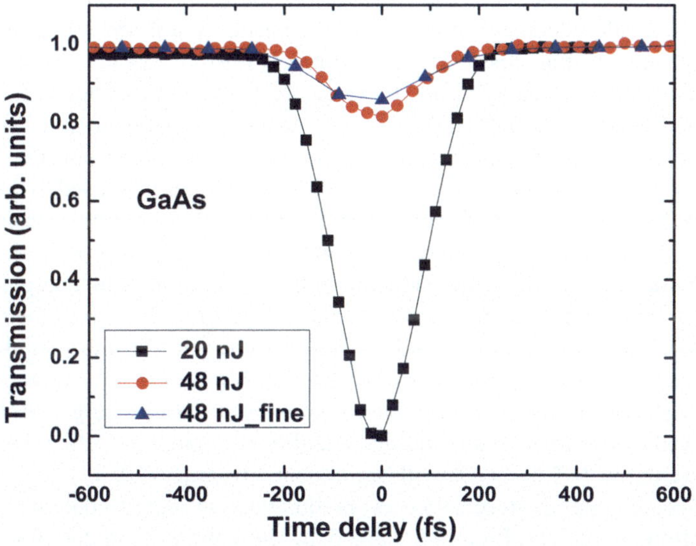

Fig. 5.7 The pump-probe data for the GaAs semiconductor at different pump beam energies represented with a 20 nJ black rectangle data point curve, 48 nJ with a red circle data point curve, and 48 nJ_{fine} with a blue triangle data point curve

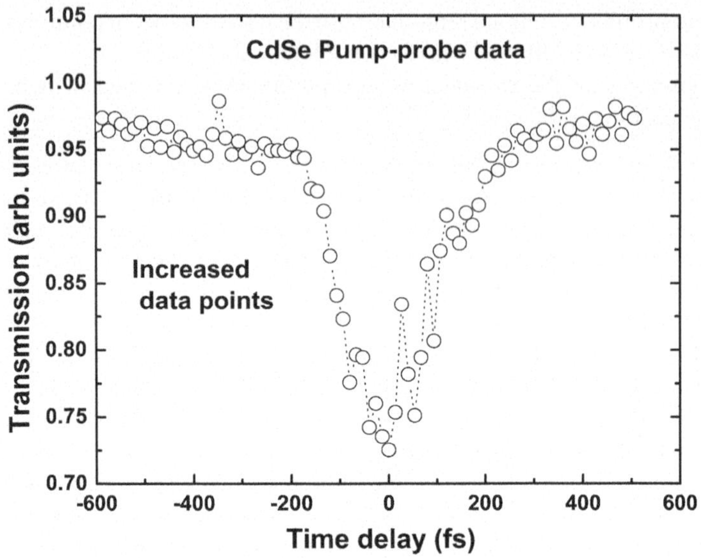

Fig. 5.8 The pump-probe data for the CdSe semiconductor where transmission is represented in arbitrary units and the time delay is shown in femtoseconds

Further, the highest value of transmission is 0.99, and then the transmission value gradually starts to decrease. The value of the transmission starts to decrease from 0.99 to around 0.93 between the time delay of −400 fs and −300 fs.

Further, a sudden decrease in the value of transmission in the range of −150 fs to 0 fs. The value of transmission is approximately 0.91 at −145 fs, drops to 0.87 at −120, and further dropped to 0.7 at 0 fs. Thus, at 0 fs the transmission is at its lowest for CdSe, so the 2PA dominates the effects of transmission in this region. Further, the value of transmission shows an increase in the transmission value from 50 fs to 600 fs. The value of transmission is approx 0.83 at 30 fs and further increased to 0.98 at 400 fs, followed by a mixed trend of value lower as compared to its initial transmission is achieved within the time delay range from 100 to 500 fs. The pump-probe data for the CdSe is similar to the GaAs semiconductor with a 20 nJ pump beam energy, showing a significant change in transmission over time delay. Table 5.1 shows the details of pump-probe data with maximum and minimum values of transmission along with the time delay values for GaAs and CdSe material. Mn-doped ZnSe QDs have been extensively studied in the literature due to their lower toxicity, larger Stokes shift, and enhanced environmental and thermal stability compared to CdSe QDs, making them an alternative to CdSe QDs with significant advantages [9, 12].

Here, Z-scan open-aperture data (OA) is illustrated in order to obtain the nonlinear absorption coefficient. Figures 5.9 and 5.10 show the Z-Scan data for the ZnSe semiconductor, where OA experimental data is shown with circles, and the red curve shows the theoretical fitting for experimental data. The experimental data is fitted

Table 5.1 Pump-probe data with maximum and minimum values of transmission along with the time delay values for GaAs and CdSe material

Sr no.	Material	Pump beam energy (nJ)	Max. value of transmission with time delay	Min. value of transmission with time delay
1.	GaAs	20 nJ	0.99 at (−500 to −300) fs	0 at 0 fs
2.	GaAs	48 nJ	0.99 at 450 fs	0.85 at 0 fs
3.	GaAs	48_{fine} nJ	1.00 at 500 fs	0.81 at 0 fs
4.	CdSe	10 nJ	0.98 at -350 fs	0.7 at 5 fs

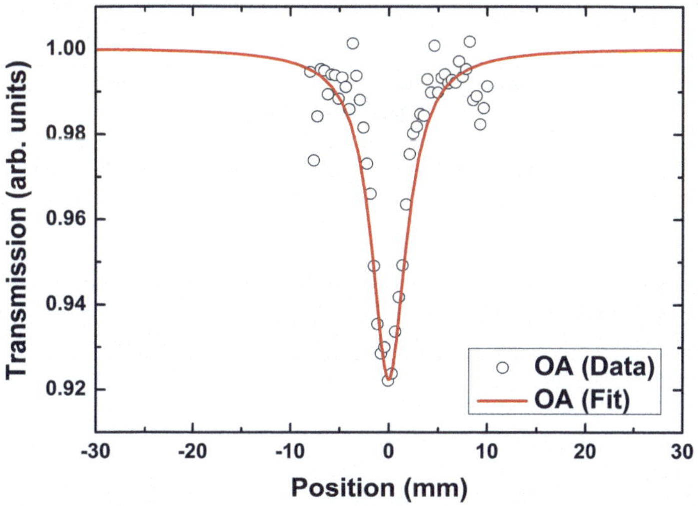

Fig. 5.9 Z-Scan data for empty arm of the ZnSe semiconductor where open-aperture (OA) experimental data is shown with circles, and the red curve shows the theoretical fitting for experimental data

by using the computational software MATLAB, which is used to determine the nonlinear absorption coefficient by inserting input parameters for the program. In order to obtain nonlinear absorption coefficient values, normalized transmission data is required. In the case of NLA measurements, the aperture is removed so that the entire beam propagates to the detector. The main cause of NLA is due to two-photon absorption or multiphoton absorption processes. 2PA in semiconductors is determined from the transition rate, and second-order perturbation theory provides us with the information related to the rate of transition from the valence to the conduction band. The use of a band-structure model is desirable, and nondegenerate 2PA can be used to determine the nonlinear refractive index through a nonlinear Kramers-Kronig transformation. Table 5.2 shows Z-Scan OA theoretically fitted parameters

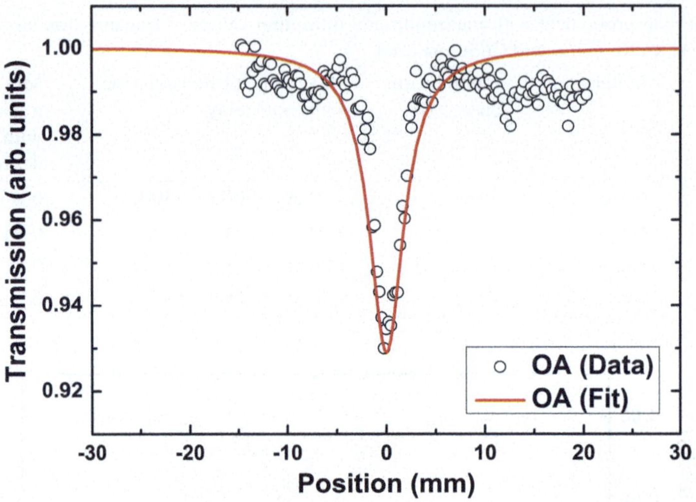

Fig. 5.10 Z-Scan data for full arm of the ZnSe semiconductor where open-aperture (OA) experimental data is shown with circles, and the red curve shows the theoretical fitting for experimental data

Table 5.2 Z-Scan OA theoretical fit of experimental data for various parameters for the empty and full arm ZnSe semiconductor

Sr no.	Parameters	$ZnSe_{Empty}$	$ZnSe_{Full}$
1.	Wavelength (nm)	800 nm	800 nm
2.	Pulse width (FWHM)	150 fs	150 fs
3.	Normalized value (CA/OA)	0.99	0.98
4.	NLA (α_2)	2.2 cm/GW	1.99 cm/GW
5.	Irradiance (I_0)	2.270 GW/cm^2	2.270 GW/cm^2

to determine Normalized value (CA/OA) and NLA (α_2) for the empty and full arm ZnSe semiconductor.

With the help of above mentioned fitted parameters, we can conclude that the value of NLA is high for ZnSe, showing that 2PA effects dominate as compared to the transmission value. So, ZnSe can be used for different types of applications such as for calibration in experiments, optical detectors, and limiters. 2PA has various applications, such as optical power limiting, 2P fluorescence imaging, 2P photodynamic cancer therapy, and 2P microfabrication [2, 10]. TAS is employed to study the ultrafast carrier dynamics, relaxation, and excitation dynamics of photocarriers in different semiconductors. Understanding the carrier dynamics in semiconductors is important for the development of future optoelectronic systems.

We investigated the carrier dynamics in different semiconductors with the help of the pump-probe technique. The decay kinetics of transient absorption for various semiconductors show broad excited state absorption in the visible region, which may be due to trapped carriers. In biological studies, they contribute to understanding the dynamics of photosynthesis, DNA photodamage, and protein dynamics. Additionally, these measurements have implications in fields like quantum optics, nanoscience, and nonlinear optics, where controlling and manipulating light-matter interactions on ultrafast timescales is crucial and opens various new ways for next-generation technologies with enhanced performance. The nonlinear optical experiments depend on both the spatial and temporal resolution of the laser beam utilized. To calculate the nonlinearity of a particular material, it is necessary to know its linear optical properties, such as the linear absorption coefficient and linear index of refraction. As systems become smaller, processes occur at a faster rate, so making optical measurements is necessary to explore the properties of semiconductors in the nanometer and femtosecond range. The study of nonlinear optical dynamics for a material opens up ways for various applications such as telecommunications, switching, data processing, sensing, quantum information processing, and imaging.

5.5 Summary

In summary, we have reviewed the materials having high value of the third-order nonlinear susceptibility and their applications for a variety of optical devices with particular attention to the various types of AOS techniques based on different physical principles. The END-2PA coefficient values in the ZnSe sample are enhanced about 270 times compared to the D-2PA and ND-2PA coefficient values for the particular selection of the wavelengths. This type of selection of wavelengths is also proposed for obtaining high NLR values of the NLO materials. We try to address the issue that why extremely nondegenerate two-photon absorption (END-2PA) coefficient values are significantly enhanced in comparison to the degenerate two-photon absorption (D-2PA) and nondegenerate two-photon absorption (ND-2PA) coefficient values for a particular selection of the wavelengths. Although investigation of the 2PA coefficient values of the semiconductor materials contributes to reduce the FOM values, the idea can be easily borrowed to enhance the NLR values by the smart selection of incident wavelengths to avoid the losses.

References

1. C.M. Cirloganu, L.A. Padilha, D.A. Fishman, S. Webster, D.J. Hagan, E.W. Van Stryland, Extremely nondegenerate two-photon absorption in direct-gap semiconductors. Opt. Exp. **19**(23), 22951–22960 (2011)
2. W. Denk, J.K. Strickler, W.W. Webb, Two-photon laser scanning fluorescence microscopy. Science **248**(4951), 73–76 (1990)
3. Y. Fainman, J. Ma, S.H. Lee, Non-linear optical materials and applications. Mater. Sci. Rep. **9**(2–3), 53–139 (1993)

4. D.A. Fishman, C.M. Cirloganu, S. Webster, L.A. Padilha, M. Monroe, D.J. Hagan, E.W. Van Stryland, Sensitive mid-infrared detection in wide-bandgap semiconductors using extreme non-degenerate two-photon absorption. Nat. Photon. **5**(9), 561–565 (2011)
5. A. González-Tudela, A. Reiserer, J.J. García-Ripoll, F.J. García-Vidal, Light–matter interactions in quantum nanophotonic devices. Nat. Rev. Phys. **6**(3), 166–179 (2024)
6. D.C. Hutchings, E.W. Van Stryland, Nondegenerate two-photon absorption in zinc blende semiconductors. JOSA B **9**(11), 2065–2074 (1992)
7. Y. Li, Q. Wu, Z. Lin, Y. Liu, Y. Zhou, Linear and nonlinear record high optical birefringence in anisotropic van der Waals materials. Light: Sci. Appl. **13**(1), 16 (2024)
8. H.S. Pattanaik, M. Reichert, J.B. Khurgin, D.J. Hagan, E.W. Van Stryland, Enhancement of two-photon absorption in quantum wells for extremely nondegenerate photon pairs. IEEE J. Quantum Electron. **52**(3), 1–14 (2016)
9. B.B. Srivastava, S. Jana, N.S. Karan, S. Paria, N.R. Jana, D.D. Sarma, N. Pradhan, Highly luminescent MN-doped ZNS nanocrystals: gram-scale synthesis. J. Phys. Chem. Lett. **1**(9), 1454–1458 (2010)
10. L.W. Tutt, T.F. Boggess, A review of optical limiting mechanisms and devices using organics, fullerenes, semiconductors and other materials. Prog. Quantum Electron. **17**(4), 299–338 (1993)
11. E.W. Van Stryland, Y.-Y. Wu, D.J. Hagan, M.J. Soileau, K. Mansour, Optical limiting with semiconductors. JOSA B **5**(9), 1980–1988 (1988)
12. P. Zhao, M. Reichert, S. Benis, D.J. Hagan, E.W. Van Stryland, Temporal and polarization dependence of the nonlinear optical response of solvents. Optica **5**(5), 583–594 (2018)

Chapter 6
Spin-Based Dichroism and All-Optical Switching in Nanocrystal Quantum Dots

6.1 Introduction to Spin-Based Dichroism in Nanocrystal Quantum Dots

The rapid growth of the current global internet with massive data communication has caused an increase in data traffic. So that our information society will inevitably demand terabit information processing or ultrafast switching. While optical fibers can theoretically carry over 50 terabits per second, current optical communication systems are limited to 10 gigabits per second due to limitations in the switching time. All-optical switches have been studied in order to overcome the switching time limit of the electro-optic switches [2]. Although organic compounds have large optical nonlinearities, semiconductors are among the most widely investigated materials because of their potential, i.e., the optical volume and the wavelength can be controlled easily by using the material size and the band gap. A saturable absorber scheme has often been tried, but the slow recovery from absorption bleaching broadens the switching time resolution and eventually hinders high-bit-rate processing. Recently, spin-based optical dichroism was found to be free from this constraint of the carrier population, and a switching time resolution of tens of picoseconds was achieved in a multi-quantum-well structure [17].

Chemically synthesized nanocrystal quantum dots (NQDs) are one of the prospective candidates because of such advantages as size tunability, homogeneity, temperature insensitivity, and low preparation cost in comparison with epitaxially grown semiconductors. Although carrier trapping to dangling bonds on the surface was a major problem, surface passivation has been improved remarkably by organic or inorganic core-shell structure [5, 6]. Regarding the three-dimensional confinement effect, large optical nonlinearities are also expected in quantum dots [1, 9, 13, 14, 18]. Since NQDs of a few nanometers in size are subject to the strong confinement in the region where the exciton Bohr radius (a_B) is larger than the dot size (a_0), the fermion character of an electron-hole pair in a dot may provide stable spin coherence even at room temperature. While the spin coherence time is known to become longer as

R. Sharma, *Nonlinear Optical Materials for All-Optical Switching Applications*,
Progress in Optical Science and Photonics 33,
https://doi.org/10.1007/978-981-96-5921-0_6

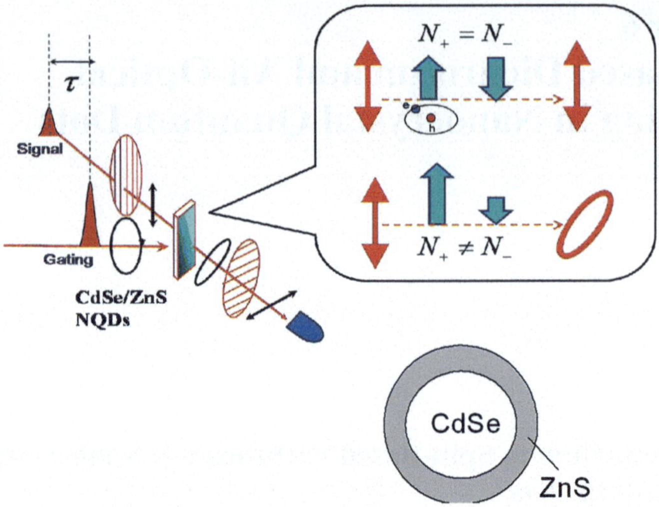

Fig. 6.1 Schematic diagram of AOS alignment and operation principle of the spin-based dichroism

the confinement increases, it also becomes shorter as the temperature increases [3]. Therefore, feasible spin states can be allowed in nanocrystals due to the balance between the temperature-induced fast spin flip and the firm spin fermion characters. In this work, we studied all-optical switching at room temperature by utilizing the degenerate spins of the ground state ($1S^e$-$1S^h_{3/2}$) and the switch-off noise in terms of spin-based dichroism.

Figure 6.1 shows the AOS operation principle schematically. The linear polarization is decomposed into two opposite circular polarizations (clockwise and anticlockwise). Each of the circular polarizations corresponds to $\pm 1\hbar$ spin angular momentum of an e-h pair (exciton). Suppose the two population are not in balance; then, the superposition of the two spin states gives rise to a phase retardation. For instance, if only the $+1\hbar$ state is fully occupied with an empty $-1\hbar$ state, the incident linear polarization turns out to be a circular polarization with a 45° phase retardation. Although selective excitation of the spin-up state ($+1\hbar$) is possible with a circular gating pulse, the two degenerate spin states gradually become balanced via the fast spin flip process. Consequently, the elliptically polarized signal beam is allowed to penetrate between a pair of perpendicular linear polarizers as long as the population imbalance is preserved. The switching time is, in fact, determined by the spin decoherence time.

In the case of strong confinement ($a_0 < a_B$), the strong spherical confinement dominates the Coulomb interaction between an electron and a hole. However, regarding a mixture of valance subbands ($J = 3/2$ and $J = 1/2$) of CdSe and the Coulomb correction, the optical transition between the electron and the hole states is identified as shown in Fig. 6.2b [6]. For a given 585 nm for the ground state ($1S^e$-$1S^h_{3/2}$), the average NQD size was estimated to be 2.97±0.21 nm (7 % size deviation), which

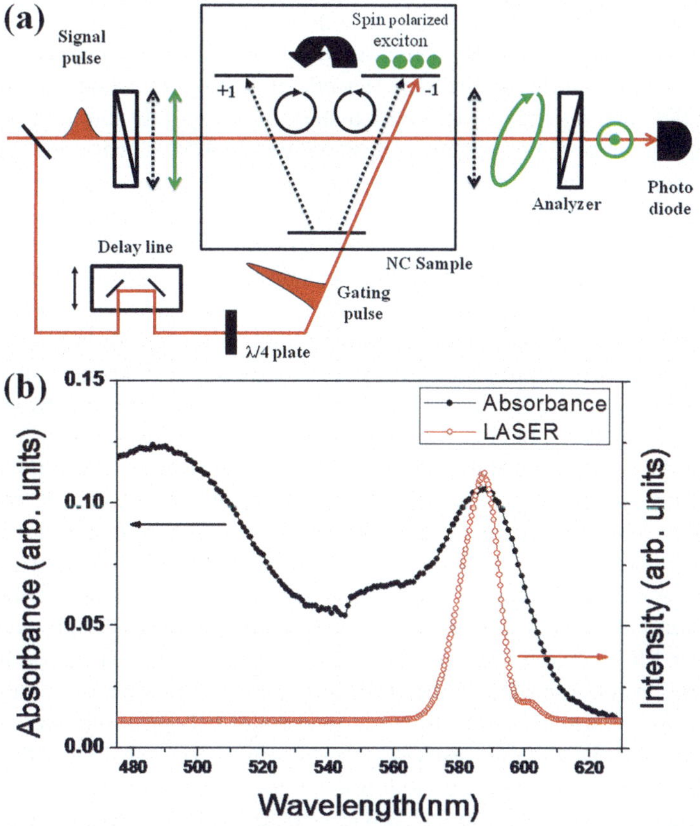

Fig. 6.2 (a) Schematic block diagram of the experimental set-up used for the pump-probe and the AOS measurements in the CdSe/ZnS NQD specimen at room temperature. (b) The absorption spectra of CdSe/ZnS NQD sample and femtosecond (fs) pulsed laser spectra centered around 585 nm are highlighted on the left and right Y-axis scales, respectively. The fs pulsed laser beam at 585 nm wavelength is obtained from a tunable optical parametric amplifier pumped by the fundamental fs laser beam at a wavelength of 800 nm with 1 kHz repetition rate [12]

is smaller than the exciton Bohr radius of 5.35 nm for CdSe [4]. ZnS shell coating enhances the surface passivation, suppressing the charge separation or surface trapping. Regarding the finite potential barrier of ZnS with respect to CdSe, the confinement energy is larger than the Coulomb energy ($\Delta E_{\text{Coulomb}}/\Delta E_{\text{conf}} \sim 0.11$).

In the last decade, considerable attention has been devoted to the development of new materials that combine ultrafast response times and high optical nonlinearities for applications in ultrafast all-optical switching (AOS) for optical communication systems. Multiple quantum wells (MQWs) and semiconductor NCs are among the most investigated materials for these applications because their optical properties can be controlled by size and high surface-to-volume ratio, which generates a large amount of surface trap states to accelerate the recovery process. It is well known that the quantum size effect leads to atomic-like optical behavior in NCs due to the quantization of energy states. By revealing the atomic-like behavior, semiconductor NC spectra exhibit a series of discrete electronic transitions between these quantized levels. On the other hand, in case of bulk materials, heavy-hole and light-hole bands are degenerate at the Γ point. However, in MQWs and NCs, these are split by the difference between the quantization energies. Since the bandgap of the heavy-hole exciton is generally lower than that of the light-hole exciton, only the heavy-hole exciton can be selectively excited by the incident light. In addition, using circularly polarized light, the spin-polarized excitons can be excited. The right (left) circularly polarized light excites the spin-up (down) excitons. Since this state is unstable, half of the spin-up excitons relax to the spin-down state, or an equilibrium state.

The time scale of spin relaxation in NCs is a few picoseconds at room temperature [3]. Therefore, sub-picosecond switching may be obtained using the spin relaxation mechanism. The spin polarization in NCs causes the circular dichroism, which vanishes at the equilibrium state. According to previous reports, it is almost evident that spin relaxation at room temperature is due to Dyakonov-Perel (DP) interaction in QWs. DP mechanism explains the spin flip processes as a result of asymmetry of the constituent atoms in the III-V compounds with the zinc-blende structure [7]. Spin-relaxation time is obtained using circularly polarized pump and probe beams.

An ultrafast all-optical switch was proposed using picoseconds spin polarization relaxation in MQW etalon structure [8]. In this device, the contrast was optimized by the polarization-rotation angle of the probe beam, and the polarization-rotation angle was shown to be proportional to the total thickness of the quantum well structure. It was predicted that the contrast could be improved by over 13 dB by optimizing the MQW etalon structure. In addition, another device was reported with low-temperature-grown Be-doped strained InGaAs/InAlAs MQWs [15]. The on/off ratios were considerably improved to 30 and 40 dB at pump pulse energies of 4 and 14 pJ, respectively. Thereafter, researchers have extended their study of AOS from MQWs to quantum dots (QDs). In this regard, there was an effort to construct an ultrafast optical switch operating at 1 Tbit/s using CdTe-quantum-dots-doped glasses [11]. Using a three-beam pump and probe experiment, it was proposed that thermal effects were responsible for a baseline in the pump and probe graphs. Pacifici et *al*. reported an all-optical modulator in which efficient interaction between two light beams at different wavelengths was achieved by converting them into co-propagating surface plasmon polaritons (SPPs) interacting by means of a thin layer of CdSe QDs [10]. The high SPP field confinement and high QD absorption cross section enable optical modulation at low-power densities in micrometer-scale planar devices.

Although all-optical switching has been extensively studied during recent years, most of the reports described above are related to MQW structures [8, 15]. However, there are limited numbers of reports in the literature that describe AOS using QDs as an investigation medium [16]. In the present chapter, we report all-optical polarization switching based on resonant excitation of spin-polarized e-h pairs in CdSe NCs. In CdSe NCs, these results suggest the potential usefulness of this optical switch for future ultrafast optical communication systems.

6.2 Experimental Scheme of All-Optical Switching

The spin dynamics can be measured by using a polarization-dependent degenerate pump-probe measurement. As the ground state ($1S^e$-$1S^h_{3/2}$) is spin degenerate with $\pm 1\hbar$ exciton spin, co-circular and cross-circular pump-probe time profiles give the populations of the remaining and the spin-flipped states under selective spin excitation, respectively. Therefore, the population difference between the spin-degenerate states can be obtained as the origin of AOS. The pump beam power was kept constant at 36 μW, which is below the absorption bleaching region. As shown in Fig. 6.3, the spin-up and -down states are balanced to 1 ps. Since the Gaussian pulse duration is 150 fs, the spin flip begins before the gating pulse injection has been completed. Interestingly, we found that incomplete convergence of the two signals is sensitive to the spatial uniformity of the NQD film. Although the NQD film was prepared by spin coating from a dense solution, spatial homogeneity control of the whole area is still difficult. The central region looks spatially homogeneous, but aggregated spotty areas are easily found near the film's edge by the naked eye. We performed identical experiments to the spatial homogeneity on various sample spots. In a comparison between the aggregated edge spot (Fig. 6.5) and the uniform central region (b), different features of convergence were seen. Regarding the ensemble effect of dense NQDs within an $\sim$ 50 μm beam size, this result suggests that NQD aggregation may affect incomplete convergence between co- and cross-circular polarizations.

Nonetheless, incomplete convergence still remains in homogeneous areas. The residual long-spin coherence due to such incomplete convergence was studied in an asymmetric quantum dot [3]. In the case of epitaxially grown quantum dots, an elongated axis is formed. For example, self-assembled QDs have a base elongated along the [110] axis, while QDs formed by interface fluctuations in narrow quantum wells are elongated along the [1$\bar{1}$0] axis. Consequently, the anisotropic exchange interaction splits the spin-degenerate $\pm 1\hbar$ states into two linearly polarized states along the principle QD axis. Although chemically synthesized NQDs are nearly

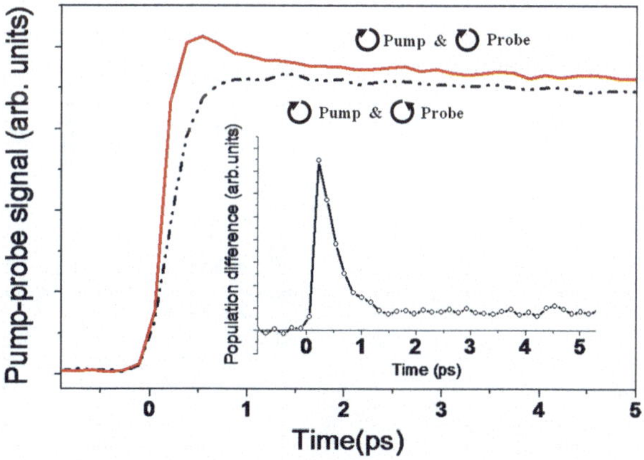

Fig. 6.3 Time evolution of pump-probe measurements of CdSe/Zns core-shell NCs at room temperature. The solid line curve shows the pump-probe signal when both pump and probe beams are in the same configuration of circular polarization, while the dotted line curve is for opposite polarization. The inset shows the difference between +1 and -1 exciton spin states

spherical, the shape asymmetry, such as prolate and oblate, is worth taking into account. Consequently, the asymmetric exchange interaction may cause slow spin decay, resulting in the incomplete convergence, while the fast convergence is governed by direct exciton spin flip. In the case of CdS QDs in a glass matrix, the slow electron spin flip with two-LO-phonon intra-level transitions is known to be driven by an exchange interaction. As shown in Fig. 6.5a, b, nearly 10% of the off-set background still appears with respect to the maximum population difference. This result is consistent with the previous result [3]. Additionally, recent work reported that the long-time accumulation of real carriers results in the background enhancement via thermal effect [11]. Since this thermal effect can be reduced by using low-frequency modulation of a mechanical chopper, this was not the case in our experiments. Consequently, we conclude that the incomplete convergence is an intrinsic phenomena in strongly confined quantum dots due to slow electron spin decay, although the spatial homogeneity of the ensemble is affected.

6.3 Noise Optimization of All-Optical Switching Signal

The pump beam is focused onto a circular spot with a 300 μm diameter in order to modify the optical response of the sample only locally in the surroundings, hence requiring a much lower energy than by illuminating the whole sample. In order to investigate the switching of the transmission spectrum, we temporally track the transmittance of the probe beam at a fixed wavelength in the time lapse of the pump pulse and, in this way, the switched spectra were obtained. The noise can be noticed in the tail of the transmitted signal in Fig. 6.4, which is due to the dominance of the background signal at low excitation power of the probe beam. The comparison of the inset of Figs. 6.3, and 6.4 shows that the noise is the dominant factor in the tail portion of the latter. This is possibly due to the presence of Glan-Taylor polarizer, having a high extinction ratio, in AOS measurements of Fig. 6.4. The purpose of temporally

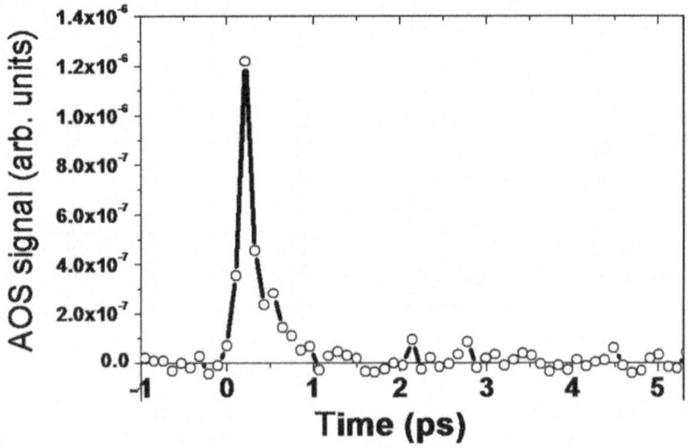

Fig. 6.4 Profile of AOS signal of CdSe/ZnS core-shell NCs at room temperature

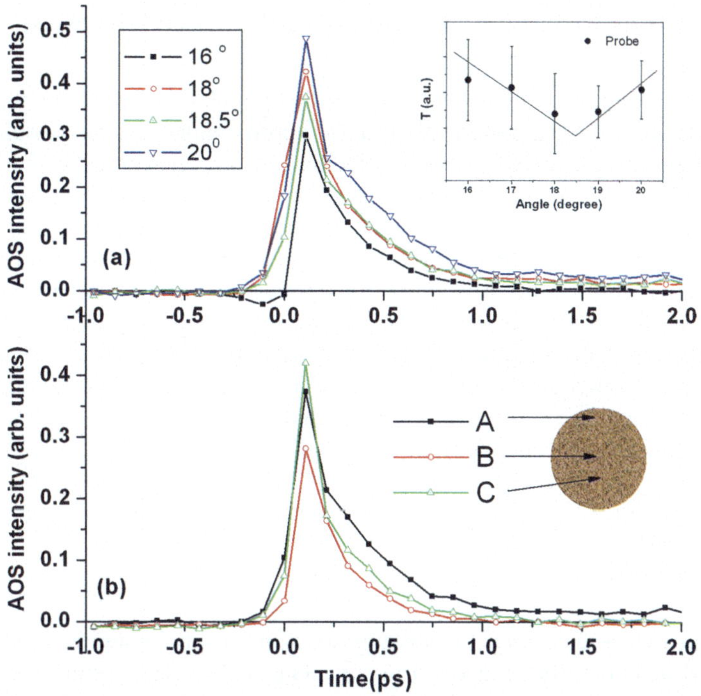

Fig. 6.5 **a** The AOS intensity is sensitive to the angle between the incident and the analyzing linear polarizer at an angle near 90°. The accurate perpendicular angle (18.5°) was found by using the signal beam modulation (inset). **b** The noise due to incomplete AOS switch-off is vulnerable to aggregation or spatial uniformity. The more central (B>C>A), the more uniform, the less aggregated

modifying the optical response of the sample is to optically generate free carriers in the NCs by pumping with an ultrafast laser pulse having a frequency around its band edge. Although this method ignores the detailed hyperfine structures of the sample under investigation [6], as reported by Klimov and co-workers, even the AOS using the spin polarization technique was described in a lucid and interesting way.

Fig. 6.5 show the AOS signal intensity utilizing spin-based dichroism. The perpendicular angle between a pair of linear polarizers, shown schematically in Fig. 6.1, is crucial for AOS operation, otherwise, leakage signals are to be detected after switch-off. A modulated probe beam, combined with a lock-in amplifier, enhances the detection accuracy, and the perpendicular angle was measured to be 18.5° (arbitrary indicator reading). The temporal profiles of the AOS signal near the perpendicular angle are shown in Fig. 6.5a. We found the AOS leakage noise to be very sensitive to the perpendicular condition, despite high extinction ratio polarizers. For instance, a 2° deviation with respect to the perpendicular condition gives rise to a 12% off-set noise. However, at this perpendicular condition, a background noise tail is still observed after 1 ps, as shown in Fig. 6.5a. This is attributed to incomplete convergence, as shown in Fig. 6.4, which results from the intrinsic slow electron spin flip.

6.4 Solutions to Various Problems and Future Direction

As the phase retardation angle due to the residual population difference ranges over a few degree, the elliptical polarization can be approximated as a linear polarization rotation. Therefore, the noise due to the incomplete convergence can be compensated for. As shown in Fig. 6.5a, the noise after 1 ps was gradually suppressed. At 16° (2.5° rotated from the perpendicular angle), the residual noise is nearly blocked while the major AOS signals are sacrificed. Figure 6.5b shows temporal profiles of the AOS signal at three different spots, considering the spatial homogeneity effect. A residual noise tail was observed at an aggregated region of spot A, but the noise was suppressed as the spatial homogeneity was improved near the central region of the sample. Consequently, the analyzer angle correction and the spatial sample homogeneity enhance the AOS time profile and may overcome the intrinsic noise due to the slow electron spin decay.

6.5 Summary

An AOS switch is discussed by utilizing exciton spin polarization in CdSe/ZnS nanocrystal quantum dots. Since a firm spin state is given as a result of an optimal balance between strong spherical confinement and fast exciton spin flip, spin-based dichroism can be utilized at room temperature. Although selective excitation of the spin-generate ground state ($1S^e$-$1S^h_{3/2}$) provides a sub-picosecond time resolution, a residual noise remains due to the intrinsic slow electron spin decay. We found that such an incomplete convergence between the spin-up and the spin-down populations also depend on spatial inhomogeneity, such as the ensemble NQD aggregation, and that the noise can be suppressed by using the analyzer angle correction and the spatial homogeneity.

References

1. C.D. Aiello, J.M. Abendroth, M. Abbas, A. Afanasev, S. Agarwal, A.S. Banerjee, D.N. Beratan, J.N. Belling, B. Berche, A. Botana, et al.: A chirality-based quantum leap. ACS Nano **16**(4), 4989–5035 (2022)
2. A.I. Ekimov, A.L. Efros, Nonlinear optics of semiconductor-doped glasses. Physica Status Solidi (b), **150**(2), 627–633 (1988)
3. A.I. Filin, P.D. Persans, K. Babocsi, M. Schmitt, W. Kiefer, V.D. Kulakovskii, N.A. Gippius, Effect of nanocrystal growth conditions on exciton decay and spin dephasing in an ensemble of cdse nanocrystals grown in glass. Phys. Rev. B-Condensed Matter Mater. Phys. **73**(12), 125322 (2006)
4. J.-H. Kim, K. Kyhm, Y. Masumoto, Ultrafast dynamics of the band-gap renormalization and the state-filling effect in strongly-confined cdse/zns nanocrystals. J. Korean Phys. Soc. **52**(1), 64–69 (2008)

5. S.M. Kim, K. Kyhm, H.-S. Yang, K.S. Hong, Optical properties and surface conditions of cdse quantum dots. J. Korean Phys. Soc. **49**(9), 688 (2006)
6. V.I. Klimov, Semiconductor and metal nanocrystals (marcel dekker inc) (2004)
7. S. Muto, A. Tackeuchi, Carrier dynamics of quantum confined structures. Mater. Sci. Eng.: R: Rep. **22**(3), 79–111 (1998)
8. Y. Nishikawa, A. Tackeuchi, M. Yamaguchi, S. Muto, O. Wada, Ultrafast all-optical spin polarization switch using quantum-well etalon. IEEE J. Sel. Top. Quantum Electron. **2**(3), 661–667 (1996)
9. K. Oshimi, H. Ishiwata, H. Nakashima, S. Mandicl', H. Kobayashi, M. Teramoto, H. Tsuji, Y. Nishibayashi, Y. Shikano, T. An et al., Bright quantum-grade fluorescent nanodiamonds. ACS Nano (2024)
10. D. Pacifici, H.J. Lezec, H.A. Atwater, All-optical modulation by plasmonic excitation of cdse quantum dots. Nat. Photon .**1**(7), 402–406 (2007)
11. L.A. Padilha, A.A.R. Neves, E. Rodriguez, C.L. Cesar, L.C. Barbosa, C.H. Cruz. Ultrafast optical switching with cdte nanocrystals in a glass matrix. Appl. Phys. Lett. **86**(16) (2005)
12. R. Sharma, J. Kim, H.-S. Yang, K. Kyhm, Sub-picosecond all-optical switching of exciton spin polarization in nanocrystal quantum dots toward terabit communication. J. Korean Phys. Soc. **55**(1), 259–262 (2009)
13. S. Shin, Y.N. Hwang, K. Chae, S.H. Park, U. Kim, Nonlinearity of the induced absorption in a cdse quantum dot in borosilicate glass. J. Korean Phys. Soc. **33**(6), 726–729 (1998)
14. E. Sutcliffe, N.P. Kazmierczak, R.G. Hadt, Ultrafast all-optical coherence of molecular electron spins in room-temperature water solution. Science **386**(6724), 888–892 (2024)
15. R. Takahashi, H. Itoh, H. Iwamura, Ultrafast high-contrast all-optical switching using spin polarization in low-temperature-grown multiple quantum wells. Appl. Phys. Lett. **77**(19), 2958–2960 (2000)
16. J. Valdmanis, R. Fork, Design considerations for a femtosecond pulse laser balancing self phase modulation, group velocity dispersion, saturable absorption, and saturable gain. IEEE J. Quantum Electron. **22**(1), 112–118 (1986)
17. M. Wraback, H. Shen, A femtosecond, polarization-sensitive optically addressed modulator based on virtual exciton effects in an anisotropically strained multiple quantum well. Appl. Phys. Lett. **76**(10), 1288–1290 (2000)
18. V. Zwiller, T. Aichele, O. Benson, Quantum optics with single quantum dot devices. New J. Phys. **6**(1), 96 (2004)

Chapter 7
Optical Nonlinearities in Organic Molecules

7.1 Introduction

The cyanine-like organic molecules were reported to show high FOM values owing to the high NLR and the small NLA. Nalwa et al. had reported third-order NLO properties which were sensitive to the Pi-conjugation length [13]. A systematic study of the relationship between the Pi-conjugation length of the model compound and its nonlinear optical response is important to understand the structural property of the NLO materials. The NLO properties of the material depend on the variation of the molecular structure of the organic materials.

7.1.1 Organic Molecules

The π electrons are delocalized in the organic molecule because π^+ and π^- bonds are alternated along the chain of chromophore. This effect not only increases the linear absorption of the organic material but also has a significant influence on increasing to third-order nonlinear susceptibility. The linear absorption wavelength can be related to the length of the chromophore chain by the following relation,

$$\lambda_{max} = \frac{8mc}{n_0}\frac{L}{N+1} \tag{7.1}$$

L and N are respectively the length of the chain and the number of π electrons. The degenerate nonlinear refraction and nonlinear absorption experimental measurements of YZ-V-69 organic molecule were performed by Hu et al., [5]. The fitting was obtained by using three-level models. The nonlinear refractive index value can be fitted by using the three-level model; however, the NLA values are slightly underestimated. This is proved from the NLR and NLA cross-sectional spectra, where the

R. Sharma, *Nonlinear Optical Materials for All-Optical Switching Applications*,
Progress in Optical Science and Photonics 33,
https://doi.org/10.1007/978-981-96-5921-0_7

negative 2PA cross section should not be confused with two-photon gain. Nevertheless, it was shown that the three-level model fitting has artifacts. Hence, they have not tried this fitting beyond 1550 nm, to avoid discrepancies between the measured and calculated values. Further, M. R. Furdinendus has used three-level SOS model to drive ND-NLR. It was reported that an intermediate state "e" between the excited state "e'" and the ground state "g" can increase the nonlinearity owing to intermediate state resonance excitation effect, as illustrated in Fig. 7.1. They have proposed that the NLR will increase for nondegenerate laser beams. They have estimated the degree of ND-NLR enhancement by applying the three-level SOS model to nondegenerate beams.

7.1.2 Various Organic Molecules with High Figure-of-Merit

Basically, the 2PA and the NLR in the organic molecules are measured by using the dual-arm Z-scan technique, as discussed in Fig. ?? of Chap. 2. In the normal course, the organic molecules' experimental studies are performed by dissolving the solute molecules in an appropriate solvent, which can create problems in measuring the nonlinearity of the solute molecules if the solvent has a significant contribution to the nonlinear signal. In the dual-arm Z-scan technique, the solution and the solvent are simultaneously scanned in the two arms of the set–up to get rid of the beam pointing instabilities, variation of the pulse width, and the spot size of the incident Gaussian beam. Consequently, the correlated solvent nonlinearity is neglected, and only the solute signal is left.

Recently, the researchers have been interested in understanding how to design the organic molecules and the materials to obtain higher FOM values. Some of the molecules are designed on the basis of the quantum chemical equation [12, 13]. The research group of S. Marder has described how to design molecules that are highly efficient for the AOS applications and are also useful for the optical limiting process. The real part of the third-order susceptibility must be higher for the AOS application; however, the imaginary part of the nonlinear susceptibility is beneficial for the optical limiting process application. It was reported that the 2PA cross section increases about 20 times when the donor molecules were attached to both ends of

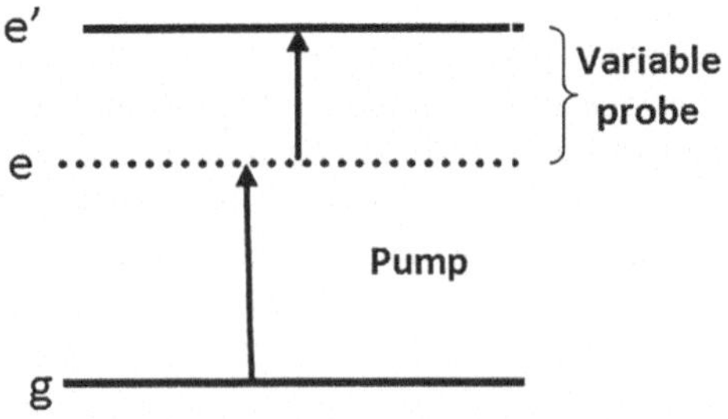

Fig. 7.1 Nondegenerate excitation of photon, where two different energy photons are used to excite the material

the stilbene molecules [12]. Recently, Zhao et al. [16] measured the NLO properties of 24 different solvents with high accuracy by using the beam-deflection technique. The beam-deflection technique allows one to deduce the electronic and nuclear contribution of the nonlinearity separately. There are mainly four types of nonlinearities present in the organic molecules and the gases: bound-electronic, nuclear collision, liberation, and diffusive reorientation. The nonlinearity due to the bound electrons is treated as an instantaneous nonlinearity, which is faster than the nuclear contribution [16]. The diffusive reorientation or the optically induced molecular reorientation can produce changes in the NLR values of the NLO organic materials. In the diffusive reorientation phenomena, the incident optical field can polarize the anisotropic molecule owing to the induced dipole moment making an angle with the incident optical field. Thus, the molecule observes a torque to align in the direction of the applied field. The liberation and the diffusive reorientation have similar origins and symmetries; however, the field-induced dipole re-radiates the electric field, which results in the NLR change via distortion of the polarizability in the neighboring molecules.

It is difficult to compare n_2 values of the materials directly because the recorded data are not taken at the same wavelength and the pulse width. For instance, in 2015, Iliopoulos et al. reported third-order NLO properties of commonly used organic solvents, which were obtained by the ps and the fs laser pulse excitation. The Z-scan measurements were carried out at 35 ps and 40 fs for the wavelength values of 532 and 800 nm, respectively; however, the optical Kerr effect (OKE) study was performed at 35 ps at the wavelength of 532 nm [7].

The third-order NLO properties of polymethines were found to be suitable for the AOS applications owing to enhanced $\pi - \pi$ interactions. Nevertheless, the thin films of these polymethines were difficult to fabricate because of the strong intermolecular forces. This problem was solved by introducing a steric bulk around π system, and the third-order nonlinear optical properties were measured by using the dual-arm Z-scan technique [3]. The Z-scan traces were obtained by our research group for two different spots on the film at a wavelength of 1550 nm which was chosen according to the telecommunication wavelengths. The 2PA and the NLR coefficients were deduced by using panels A, C and B, D of Fig. 7.2, respectively, and the FOM value of the thin film was obtained to be 7.24.

In another report, the researchers have fabricated the phosphonium-containing cationic polyelectrolyte by using a facile click-chemistry which can translate the favorable molecular properties of the cyanines into macroscopic properties of the material [2]. In order to investigate the NLO properties, different complexes of the cyanine were prepared by using dichloromethane (DCM) as solvent, and the thin films were deposited onto the fused silica substrate. The FOM of one of the samples dissolved in the DCM, PE 1-TCF1, attained a maximum value of 39 when the dual-arm Z-scan measurements were carried out at a wavelength of 1300 nm, as shown in Fig. 7.3 [10]. The high FOM results were explained owing to better dispersion of the cyanines in the best polyelectrolyte, which preserved the symmetry of the cyanines. Nevertheless, there were many other organic dyes reported by the researchers that

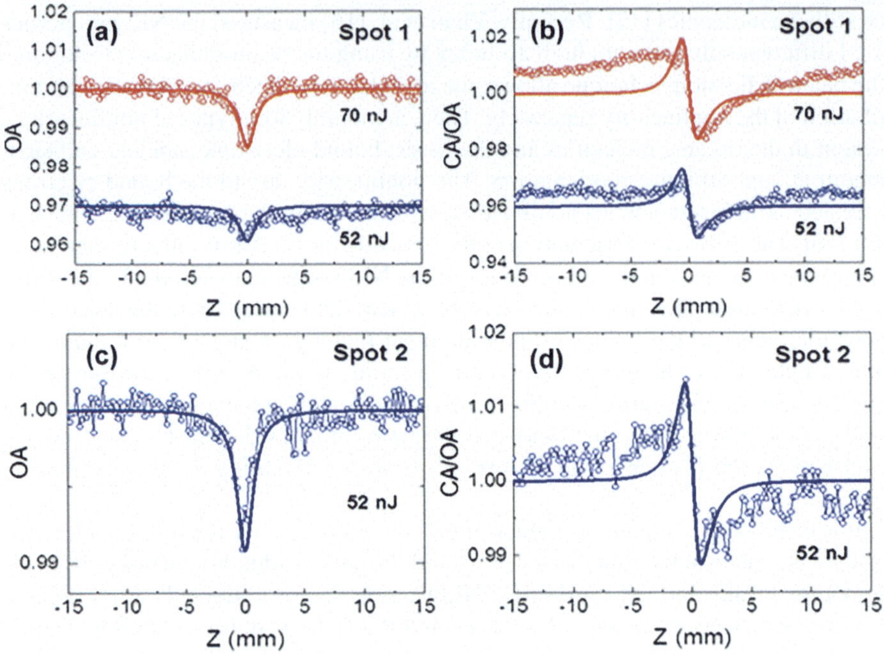

Fig. 7.2 The dual-arm Z-scan traces of 6'pd thin film recorded at a wavelength of 1550 nm. A and C are the open-aperture scans while B and D are the CA/OA scans. Reproduced with permission from [2]

can be considered for the AOS applications owing to high FOM value at the telecommunication wavelength. One of these kinds of dyes was reported by the researchers of the Georgia Institute of Technology, which can be very suitable for AOS applications. On the other hand, the FOM for 1140 nm thin film on fused silica substrate attained a maximum value of 20 at the wavelength of 1300 nm.

In order to enhance the FOM value, the 2PA should be avoided by selecting excitation photon energies smaller than the 2PA energy; however, the large detuning energy term still remains. To further reduce the detuning energy term, twice the excitation photon energy value is selected in such a way that it lies between the 2PA energy states. This concept was applied to a soluble bis(selenopyrylium) heptamethine dye where the NLR values were found to be larger over the telecommunication wavelengths starting from 1300 to 1550 nm, while the nonlinear loss was reduced by two orders of magnitude [4]. In order to estimate the FOM values of the solute in the solution, the nonlinearity in the molecules can be identified in terms of the 2PA and the NLR cross-sectional values. The solubility of the solute in the solvent molecules is also an important parameter in the FOM values, which can be beneficial to increase the concentration of the solution. Many cyanine-like molecules show relatively small FOM values owing to the presence of the large 2PA bands near the linear absorption edge; however, recently, a thiopyrylium polymethine (YZ-V-69)

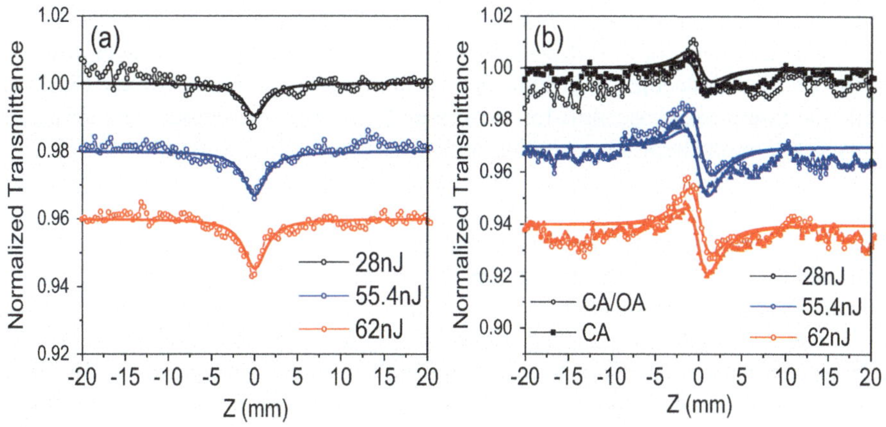

Fig. 7.3 **a** The OA dual-arm (DA) Z-scans of a-1.4-μm thick PE1-DOB-C9 neat film by using three different pulse energies, **b** the CA DA Z-scans along with the divided result of the CA signal by the OA signal with three different pulse energy measurements at a wavelength of 1.55 μm. The solid lines in (**a**) and (**b**) are theoretical fits obtained by using the Z-scan theory. Reproduced with permission from [10]

dye dissolved in carbon tetrachloride had shown FOM values > 400 near the absorption edge. Nevertheless, the value of the FOM remains about 42 and 10 at 1300 and 1550 nm, respectively. There are mainly two reasons for the high FOM values of the YZ-V-69 molecule. The first reason is the large solubility of the order of 10^{-3} M in the solvent while the second reason is attributed to the fact that only one 2PA band was observed at the main 1PA (1700 nm), but no 2PA and excited state absorption (ESA) was observed near its 1PA edge [6].

Once the suitable materials with high FOM values are identified, the next task is to demonstrate the AOS device by using a free space experimental setup. In this regard, various experimental configurations were proposed to know about the phase change provided by the NLO material; however, the Mach–Zehnder interferometry configuration can directly predict the phase change in a better way.

In the Mach–Zehnder interferometry configuration, an incident beam is split into the pump and the probe beams by using a beam splitter. The probe beam is further divided into two beams; one beam is directed onto the sample, which is further allowed to transmit through another beam splitter, while the second beam meets the beam splitter through a mirror. The interference fringes can be obtained if the spatial and temporal overlap of the two beams is taken into account. In this configuration, if the pump beam is shining onto the sample, then the central bright fringe will convert to the dark fringe if the phase change is π. Once the AOS device in the free space works, the next step would be to look for the practical communication applications that can be performed by using the slot waveguide geometry discussed below. There are also other kinds of waveguides, such as silicon-based strip and nonlinear silicon on-insulator waveguide; however, high switching speed was found in the case of silicon-organic hybrid slot waveguide. The cross-sectional view of the waveguide geometry of the silicon ribs fabricated on the SiO_2 buffer layer covered by the organic molecules with high optical nonlinearity is illustrated by Fig. 7.4a while the nonlinear phase shift for the slot and the stripe waveguide geometries of the silicon-organic hybrid approach is shown by Fig. 7.4b [1, 8, 9, 11, 14, 15].

The unstrained bulk silicon is a centrosymmetric material, therefore, the second-order nonlinear optical response is absent. Nevertheless, the third-order NLO response is present in the silicon; however, the integration into the AOS device is limited by the two-photon absorption, and the switching speed is inhibited due to the two-photon absorption-induced free-carrier absorption. The data transfer rates of 10 and 40 Gbs^{-1} in the bare silicon waveguides and the reverse biased p-i-n diode were reported. In order to obtain higher processing speeds of the AOS device, the silicon was combined with the material having a fast electronic third-order nonlinearity, which resulted in all-optical demultiplexing of up to 170.8 40 Gbs^{-1}. The details of the AOS device fabrication were reported in Ref. [8]. Here we discuss in briefly discuss the slot waveguide geometry. In the first step, the silicon-organic hybrid structure was produced on the silicon-on-insulator slot waveguide, which is further covered by a layer of the NLO organic material by using the vapor deposition techniques. A deep ultraviolet lithography technique was used at a wavelength of 193 nm on an ASML PAS5500/1100 stepper. A molecular beam was generated by heating powdered organic material to be deposited onto the surface of the substrate. The organic cladding material was deposited using the spin-coating technique. In order to fill the organic cladding material homogeneously, a small molecule, DDMEBT23 (2-[4-(dimethylamino)phenyl]-3-f[4-(dimethylamino) phenyl]ethynyl buta-1,3-diene-1,1,4,4-tetracarbonitrile) was filled by using a molecular beam deposition technique.

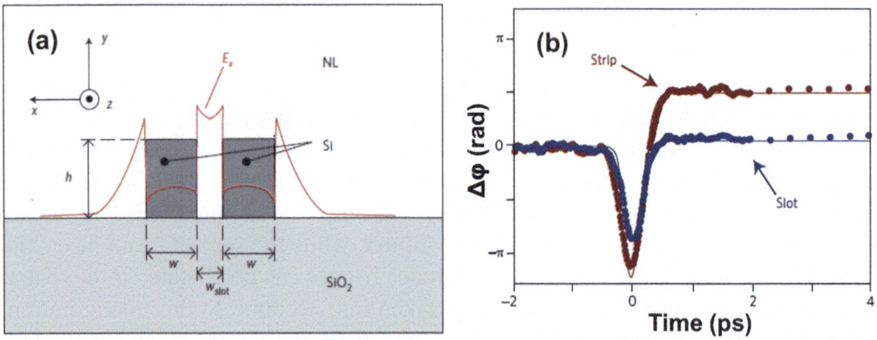

Fig. 7.4 **a** The cross-sectional view of the waveguide geometry of the silicon ribs fabricated on the SiO_2 buffer layer covered by the organic molecules which have a high value of the optical nonlinearity. **b** The comparison of the nonlinear phase shift for the slot and the stripe waveguide geometries of the silicon-organic hybrid approach. The solid circles are the experimental values, while the lines represent the fitted values. Reproduced with permission from [8]

7.2 Summary

The chapter explained the estimation of the figure-of-merit of various organic molecules, which are useful for all-optical switching applications. The question "how to estimate the nonlinear absorption and refractive index of the organic molecules" is addressed. The question "how to engineer and design the organic molecules to produce high figure-of-merit values for the all-optical switching applications" is addressed. We explored the possibility of obtaining high FOM values in the cyanine-like organic molecules by the selection of various wavelengths. We also discussed the scientific reports about the thiopyrylium polymethine (YZ-V-69) dye dissolved in the carbon tetrachloride, which can show extremely high FOM values near the absorption edge. Moreover, we extended the practical applications by proposing the incorporation of the high FOM material in a slot waveguide geometry to obtain high switching speed.

References

1. B. Cui, P. Xu, K. Fan, Y. Zhen, X. Li, R. Lu, P. Wang, X. Guo, L. Tong, Small-molecule organic ice microfibers. Sci. Adv. **11**(2), eads2538 (2025)
2. I. Davydenko, S. Barlow, R. Sharma, S. Benis, J. Simon, T.G. Allen, M.W. Cooper, V. Khrustalev, E.V. Jucov, R. Castaneda et al., Facile incorporation of pd (pph3) 2hal substituents into polymethines, merocyanines, and perylene diimides as a means of suppressing intermolecular interactions. J. Am. Chem. Soc. **138**(32), 10112–10115 (2016)
3. M.R. Ferdinandus, M. Reichert, T.R. Ensley, D.A. Fishman, S. Webster, D.J. Hagan, E.W. Van Stryland, Measuring small solute nonlinearities in solution by dual-arm z-scan technique, in *Frontiers in Optics* (Optica Publishing Group, 2012), pp. FM4I–6
4. J.M. Hales, J. Matichak, S. Barlow, S. Ohira, K. Yesudas, J.-L. Brédas, J.W. Perry, S.R. Marder, Design of polymethine dyes with large third-order optical nonlinearities and loss figures of merit. Science **327**(5972), 1485–1488 (2010)
5. H. Hu, Third order nonlinearity of organic molecules (2012)
6. H. Hu, T.R. Ensley, M. Reichert, M.R. Ferdinandus, D. Peceli, O.V. Przhonska, S.R. Marder, A.K.-Y. Jen, J.M. Hales, J.W. Perry et al., Optimization of the electronic third-order nonlinearity of cyanine-like molecules for all optical switching, in *Organic Photonic Materials and Devices XVI*, vol. 8983 (SPIE, 2014), pp. 898303
7. K. Iliopoulos, D. Potamianos, E. Kakkava, P. Aloukos, I. Orfanos, S. Couris, Ultrafast third order nonlinearities of organic solvents. Opt. Exp. **23**(19), 24171–24176 (2015)
8. C. Koos, P. Vorreau, T. Vallaitis, P. Dumon, W. Bogaerts, R. Baets, B. Esembeson, I. Biaggio, T. Michinobu, F. Diederich et al., All-optical high-speed signal processing with silicon-organic hybrid slot waveguides. Nat. Photon. **3**(4), 216–219 (2009)
9. C.-W. Lee, C. Kim, W. Yoon, H. Yun, M. Jazbinsek, F. Rotermund, O.-P. Kwon, Organic thz crystals based on off-diagonal optical nonlinearity with optimal in-plane polar axis. Adv. Opt. Mater. **12**(33), 2401590 (2024)
10. Z. Li, A.A. Syed, P. Zhao, J.C. Yang, R. Sharma, T.R. Ensley, J.D. Matichak, I. Davydenko, S.-H. Jang, D.J. Hagan et al., Cationic polyelectrolyte for anionic cyanines: an efficient way to translate molecular properties into material properties. J. Am. Chem. Soc. **141**(43), 17331–17336 (2019)
11. K. Liu, Bulk monolayer: bring monolayer optical nonlinearity into solution-processible bulk films. Matter **7**(8), 2771–2772 (2024)

12. S.R. Marder, Materials for third-order nonlinear optics. MRS Bull. **41**(1), 53–62 (2016)
13. Hari Singh Nalwa, Organic materials for third-order nonlinear optics. Adv. Mater. **5**(5), 341–358 (1993)
14. C. Valverde, F.A.P. Osório, Study of nonlinear optical properties of a self-healing organic crystal. ACS omega **9**(36), 38295–38302 (2024)
15. Z. Wang, W. Han, B. Fu, H. Kang, P. Cheng, J. Guan, Y. Zheng, R. Shi, J. Xu, X.-H. Bu, Mechanical twisting-induced enhancement of second-order optical nonlinearity in a flexible molecular crystal. J. Am. Chem. Soc. (2025)
16. P. Zhao, M. Reichert, S. Benis, D.J. Hagan, E.W. Van Stryland, Temporal and polarization dependence of the nonlinear optical response of solvents. Optica **5**(5), 83–594 (2018)

Chapter 8
All-Optical Switching in High Figure-of-Merit Materials

8.1 Overview of All-Optical Switching and Its Potential Applications

High figure-of-merit materials are crucial for the advancement of all-optical switching, enabling faster and more efficient data processing in optical communication systems. The examples discussed in this article, including chalcogenide glasses, silicon photonic crystals, quantum dots, organic nonlinear optical materials, and graphene, highlight the diverse range of materials that possess the desired properties for all-optical switching applications. Continued research and development in these materials, as well as exploration of new materials, hold great promise for achieving further breakthroughs in the field of all-optical switching, leading to the next generation of high-speed and energy-efficient optical communication systems. The all-optical-switching (AOS) is described for the nanocrystal quantum dots (NQDs) and two-dimensional (2D) materials by Fig. 8.1, which is followed by the discussion of the low figure-of-merit (FOM). In order to enhance the FOM values in the organic materials, the idea of smart selection of the wavelengths is borrowed from extremely nondegenerate two-photon absorption (END-2PA) coefficient values in the semiconductors. The organic material is incorporated in a slot wave guide geometry for telecommunication applications.

8.2 Brief History of All-Optical Switching Research

The basic principle of the all-optical switching device based on the change in direction of the output beam in the presence of the switching beam is highlighted in Fig. 8.2. The situation is highlighted in Fig. 8.2a where the input beam is incident on a nonlinear optical material and the output beam is emitted in a certain direction. This state is referred to as a switch-off state of the AOS device. On the other hand, when input

R. Sharma, *Nonlinear Optical Materials for All-Optical Switching Applications*,
Progress in Optical Science and Photonics 33,
https://doi.org/10.1007/978-981-96-5921-0_8

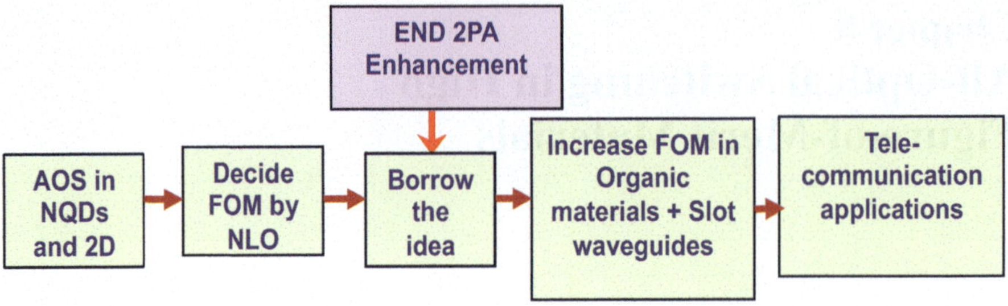

Fig. 8.1 Schematic block diagram depicts an overview of the main idea of the present investigations

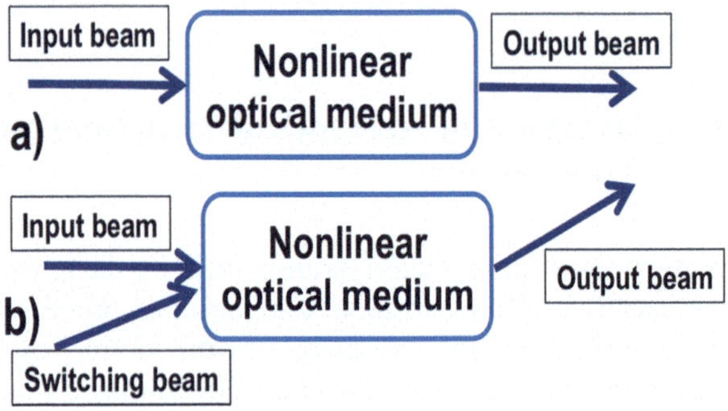

Fig. 8.2 Basic principle of a directional all-optical switching device **a** when only input beam is present and **b** when input as well as the switching beams are present

as well as the switching beams are incident on the nonlinear material, the output beam changes its direction, as illustrated by Fig. 8.2b. This state is referred to as a switch-on state of the AOS device. Note that the switch-on and switch-off states in a particular switching mechanism are arbitrary, and the nomenclature may vary in different books. The basic principle of the all-optical switching device based on the absorptive nature in the presence of the switching beam is highlighted in Fig. 8.3. Here, the situation is highlighted in Fig. 8.3a where the input beam is incident on a nonlinear optical material and the output beam is emitted. This state is referred to as a switch-off state of the AOS device. On the other hand, when input as well as the switching beams are incident on the nonlinear material, there is no emission of the output beam, as illustrated by Fig. 8.3b. This state is referred to as a switch-on state of the AOS device.

8.2.1 Examples of High Figure-of-Merit Materials for All-Optical Switching

In the field of optics, all-optical switching plays an important role in enabling faster and more efficient communication systems. All-optical switches are capable of manipulating light signals without the need for electro-optic conversions, allowing

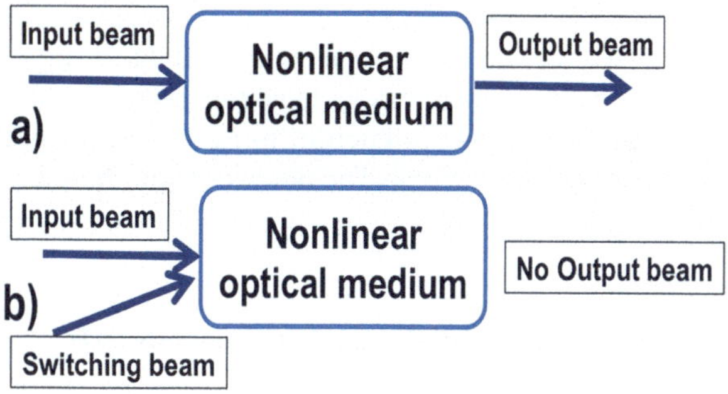

Fig. 8.3 Basic principle of an absorptive all-optical switching device **a** when only input beam is present and **b** when input as well as the switching beams are present

for high-speed data transmission and processing. One critical aspect of all-optical switches is the choice of materials with high figure-of-merit (FOM). These materials possess unique properties that make them ideal candidates for achieving efficient and reliable all-optical switching. Some remarkable examples of high figure-of-merit materials are discussed that have revolutionized the field of all-optical switching.

8.2.1.1 Chalcogenide Glasses

Chalcogenide glasses, composed of sulfur, selenium, and tellurium, have emerged as promising materials for all-optical switching. These glasses exhibit high nonlinearity due to their unique structure, which can be used for strong light-matter interactions. Chalcogenide glasses are capable of ultrafast response times and possess excellent transparency in the mid-infrared region. Moreover, their compatibility with optical fibers makes them highly suitable for integration into existing communication systems.

8.2.1.2 Silicon Photonic Crystals

Silicon photonic crystals are another notable class of materials for all-optical switching applications. These crystals exhibit a photonic bandgap structure that allows for the control and manipulation of light propagation. By engineering the lattice structure, it is possible to create defect states within the bandgap, enabling efficient light modulation and switching. Silicon photonic crystals offer low insertion losses, high extinction ratios, and compact device footprints, which make them attractive for on-chip integration.

8.2.1.3 Quantum Dot

Quantum dots are nanoscale semiconductor particles that exhibit unique optoelectronic properties. Quantum dots can confine electrons and holes in three dimensions, leading to discrete energy levels and size-tunable absorption and emission spectra. This characteristic allows for efficient light absorption and subsequent switching with low-power consumption. Quantum dots have shown great potential for achieving ultrafast, all-optical switching in various wavelength ranges, including the visible and near-infrared regions. The details of spin-based dichroism all-optical switching mechanism in the nanocrystal quantum dots are discussed at length in Chap. 6.

8.2.1.4 Organic Nonlinear Optical Materials

Organic nonlinear optical (NLO) materials are of significant interest in the scientific community for their exceptional nonlinear optical properties. These materials possess large hyperpolarizabilities, allowing for strong nonlinear responses to light signals. NLO materials can exhibit high figure-of-merit values due to their enhanced susceptibility to light-induced changes. Organic NLO materials can be engineered at the molecular level, enabling fine-tuning of their properties to meet specific requirements in all-optical switching applications.

8.2.1.5 Graphene

Graphene, a two-dimensional carbon allotrope, has attracted immense interest in the field of photonics. It possesses unique optical and electronic properties, making it a promising material for all-optical switching. Graphene exhibits ultrafast carrier dynamics and an exceptional Kerr nonlinearity, allowing for efficient light modulation and ultrafast response times. Its compatibility with silicon-based integrated circuits makes it a strong candidate for future optoelectronic devices.

8.2.1.6 Semiconductor Materials

Semiconductor phosphor material produces the best results in terms of increased luminescence quantum yield when doped with certain impurities. Nevertheless, the investigation of the third-order nonlinear optical susceptibility of the phosphor materials can be exploited for various switching applications. In this regard, the nonlinear optical properties of semiconductor phosphors are explored, where the knowledge of absorption and refraction is utilized in various optical and detector applications.

8.3 Pump-Probe Measurements and Temporal Walk-Off

The nondegenerate two-photon absorption of a ZnS single crystal sample is measured using the experimental configuration illustrated in Fig. 8.4a, where the details

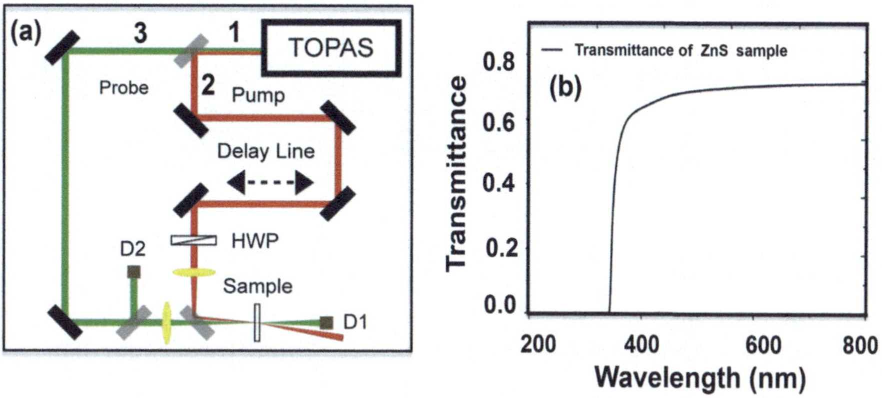

Fig. 8.4 **a** The experimental configuration of pump-probe measurements involves a femtosecond pulsed laser beam (beam 1) from the TOPAS regenerative amplifier, which is split into two parts: beam 2 and beam 3. Beam 2 and beam 3 are respectively named the pump and probe beams. Laser beam 2 is directed onto the sample through an optical delay line and a half-wave plate. Conversely, the laser probe beam 3 is directed onto the sample with the help of a pair of mirrors. A portion of the probe beam is also directed toward detector D2. The pump and probe laser beams are directed onto the ZnS sample, and changes in the probe beam are measured using detector D1. **b** Transmittance spectra of the ZnS single crystal sample [5]

of various components are highlighted in the caption. The transmission spectra illustrated in Fig. 8.4b show two-photon absorption because the energies of the pump and probe photons are smaller than the energy bandgap of the ZnS material [7, 27]. A problem arose with chirp and temporal walk-off of the laser pulses when a white light continuum was used as the probe beam. This issue was resolved by applying a correction to the two-photon absorption cross section. The cross-correlation signals of ZnS were estimated at probe wavelengths of 480, 532, and 580 nm, with the pump beam energy and beam width being 5 μJ and 500 μm (HW1/e^2M), respectively. The transmittance and temporal width of the signal depend on the relative velocity of the incident pulses onto the sample and the value of the nonlinear optical susceptibility. It is important to note that the length of interaction of the pump and probe beams is significantly reduced due to the group velocity mismatch and different wavelengths in the white light continuum. This walk-off effect resulted in a decrease in transmittance when the difference between the frequencies of the pump and probe was increased.

The results for the nondegenerate two-photon absorption coefficient (β) in ZnS are represented in Fig. 8.5(B), with theoretical estimations included according to Refs [3, 4, 10]. The group velocity mismatch (ρ) and nonlinearity (τ) parameters of the material corresponding to the wavelengths were 16.4 and 4.5 at 480 nm, 9.9 and 1.9 at 532 nm, and 6.2 and 0.6 at 580 nm, as shown in Fig. 8.5(A). This indicates that a significant correction is required to estimate the degenerate and nondegenerate cross-sectional values when the pump and probe beams have a large separation of frequencies.

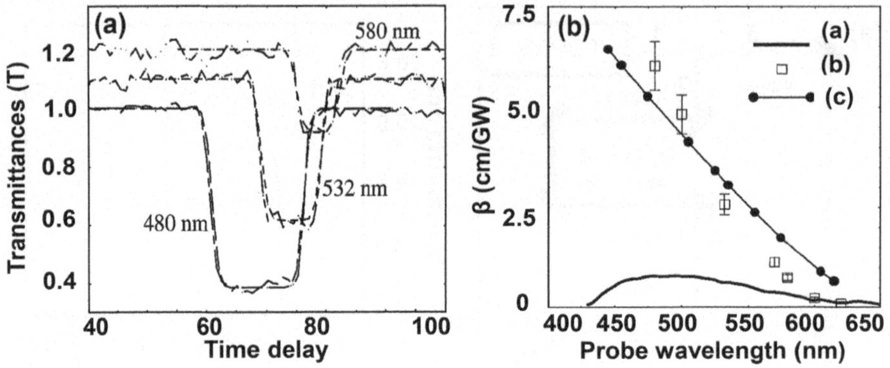

Fig. 8.5 (A) Transmittance as a function of the time delay for a ZnS sample at wavelengths of 480, 532, and 580 nm. The experimental results (dashed line) are fitted with a solid line curve whose details are referred to in the text. (B) Nondegenerate two-photon absorption coefficient as a function of probe wavelength for the ZnS sample **a** white light continuum spectral measurement without including pump-probe walk-off. **b** pump-probe walk-off correction is applied **c** calculated spectra of ZnS using theoretical model. Reprinted with permission with [26] © The Optical Society

The transmittance can be estimated by fitting the required parameters using the following Eq. 8.1:

$$T(\sigma, \tau_d, W, \rho, \tau) = \frac{exp(-2\sigma)}{(W\sqrt{\pi})}\int_{-\infty}^{+\infty} exp\{-(\frac{\tau+\tau_d-\rho}{w})^2 - \frac{\Gamma\sqrt{\pi}}{\rho}[erf(\tau) - erf(\tau-\rho)]\}d\tau \tag{8.1}$$

Here α, τ_d, W are the linear absorption coefficient, temporal walk-off parameter, temporal width of the probe and pump pulses, respectively. If n, λ, c are the refractive index at wavelength λ and speed of light, respectively, then the nonlinearity can be estimated by following Eq. 8.2:

$$\Gamma = L(n/n_p)I_p{}^0\beta \tag{8.2}$$

Here, L and β are respectively the sample thickness and the nondegenerate two-photon absorption coefficient. n and n_p are the linear refractive indices for the pump and probe beams, respectively.

Further, Fig. 8.6 depicts the nondegenerate measurement results where a white light continuum probe was used at λ_p=1.2 μm along with the degenerate two-photon absorption spectrum [13, 28]. The white light continuum pump-probe technique is beneficial to obtain a database of spectra of various organic dyes, which can be applied to investigate structure-property relations [38].

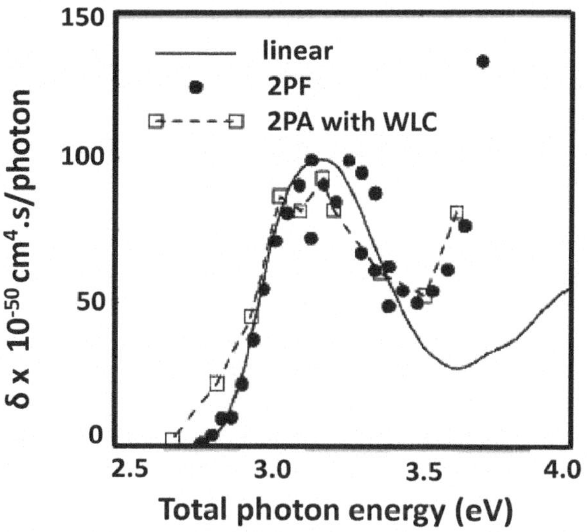

Fig. 8.6 Degenerate two-photon absorption cross section is represented with the help of solid circles, where two-photon fluorescence is taken into account. Nondegenerate cross section is represented by using open square symbols. The solid line represents the linear absorption spectrum. Reprinted with permission with [26]© The Optical Society

8.4 Properties and Characteristics of High Figure-of-Merit Materials

The third-order nonlinear optical susceptibility is the sum of the real and the imaginary parts and is given by the following equation

$$\chi^3 = Re(\chi^{(3)}) + Im(\chi^{(3)}) \tag{8.3}$$

The real part of the nonlinear optical susceptibility can be related to the NLR values by relation

$$Re(\chi^{(3)}) = 2{n_0}^2\epsilon_0 c n_2 \tag{8.4}$$

The imaginary part corresponds to the NLA,

$$Im(\chi^{(3)}) = \frac{{n_0}^2\epsilon_0 c^2\alpha_2}{\omega} \tag{8.5}$$

The NLR and the NLA coefficients of the materials can be measured by the above-listed equations, which can be further used to decide the figure-of-merit (FOM) of the NLO material. The FOM is a useful quantity for the AOS applications and can be written as

$$FOM = |\frac{Re(\chi^{(3)})}{Im(\chi^{(3)})}| \tag{8.6}$$

$$FOM = \frac{4\pi n_2}{\lambda_0\alpha_2} \tag{8.7}$$

where λ_0 is the wavelength of the incident beam. But we cannot measure the transient NLR and NLA effects in materials by using single and dual-arm Z-scan. Hence, the beam-deflection technique was used to determine n_2 values with the $\lambda/20,000$ accuracy [1, 11, 37].

In order to fabricate the high figure-of-merit materials, one of the methods was highlighted by using the selenopyrylium end group to increase the conjugation length of the molecule design of the pi-conjugated organic materials without any change of the symmetry of the bond length. This type of design of the pi-conjugated organic materials, along with the selection of the photon energy to reduce the linear and nonlinear losses, can significantly enhance the FOM values. The details of the fabrication process are referred to in Refs [12, 14]. On the other hand, the chemical precipitation method was used to synthesize the NQDs, where the CdO, the hexadecylamine, the Se powder, the trioctylphosphine oxide, the trioctylphosphine, and the stearic acid were used to prepare the precursor solutions. The surface of the CdSe was passivated by alkylamine ligands to obtain a high quantum yield. The precursors of CdO and stearic acid were heated in a flask to 200 °0C and cooled down to room temperature. Thereafter, the trioctylphosphine oxide and the hexadecylamine were added, and the solution was again heated. The Se solution and the trioctylphosphine were mixed together and injected into the solution prepared in the previous step [20].

8.5 Explanation of Different Mechanisms of All-Optical Switching

All-optical switching encompasses various mechanisms that enable the control and manipulation of light signals without the need for electrical conversion. These mechanisms include nonlinear effects such as the Kerr effect, two-photon absorption, and cross-phase modulation. The Kerr effect exploits the nonlinear response of materials to intense light, leading to changes in refractive index and subsequent modulation of transmitted light. Two-photon absorption occurs when two photons are simultaneously absorbed, triggering changes in the material's properties. Cross-phase modulation, on the other hand, utilizes the refractive index changes induced by the interaction of two light beams of different intensities.

High figure-of-merit materials are essential for achieving efficient and effective all-optical switching. These materials possess unique properties that enhance their nonlinear responses and enable robust switching capabilities. High FOM materials exhibit a large nonlinear refractive index change and low optical losses, ensuring minimal signal degradation during switching operations. Additionally, they offer fast response times and low-power consumption, making them ideal for high-speed communication systems. The careful selection and design of high FOM materials are crucial to achieving optimal all-optical switching performance. A comparative analysis of different all-optical switching mechanisms reveals their distinctive characteristics and benefits. The Kerr effect, being a ubiquitous mechanism, offers fast response times and compatibility with various materials. Two-photon absorption, while requiring higher light intensities, provides greater control and precision due to its threshold-based behavior. Cross-phase modulation enables efficient wavelength conversion and signal processing, making it suitable for advanced optical networks. Each mechanism has its strengths and limitations, and the choice depends on the specific requirements of the application.

Experimental investigations play a crucial role in understanding and advancing the field of all-optical switching. Various techniques are employed to study the behavior and performance of all-optical switches. Nonlinear optical characterization methods, such as pump-probe spectroscopy and Z-scan techniques, allow for the measurement of material properties and response times. Time-resolved experiments enable the observation of ultrafast switching dynamics. Advanced setups involving mode-locked lasers and pulse shaping techniques provide precise control over light parameters, facilitating detailed investigations into all-optical switching phenomena. Recent experimental studies have demonstrated the effectiveness of high FOM materials in achieving all-optical switching. For instance, chalcogenide glasses have been extensively investigated for their excellent nonlinear properties and compatibility with fiber-optic systems. Silicon photonic crystals have been explored for their precise control over light propagation and efficient switching functionalities. Quantum dots have shown promise in achieving ultrafast and low-power switching in the visible and near-infrared regions. Organic nonlinear optical materials have been studied for their large hyperpolarizabilities and tunable properties. These examples highlight the significant progress made in utilizing high FOM materials for all-optical switching applications.

8.6 Photonic Switching Applications

Since the reduction in particle size can affect the nonlinear properties, semiconductor quantum dots and photonic cavities can be prospective candidates for photonic switching applications. Semiconductor quantum dots are suitable for photonic switching applications, although there are various challenges highlighted in Fig. 8.7. In order to attain a reasonable signal-to-noise ratio in a particular switch for practical applications, the number of incident photons is limited to one thousand [26, 33]. On the other hand, the dashed line proposed by Wada in 2004 shows a significant step toward achieving improved performance of photonic switches based on all-optical, electronic, and optoelectronic approaches. In order to develop the practical logic devices, the optical power consumption and the cross section of the gain medium must be kept small [18]. The requirement for a photonic switch is that it should convert electrical signals to optical signals with a recovery time on the order of picoseconds or femtoseconds. Such photonic switches may have terahertz switching speeds with minimum loss optical waveguides, providing high switching speeds at a low cost.

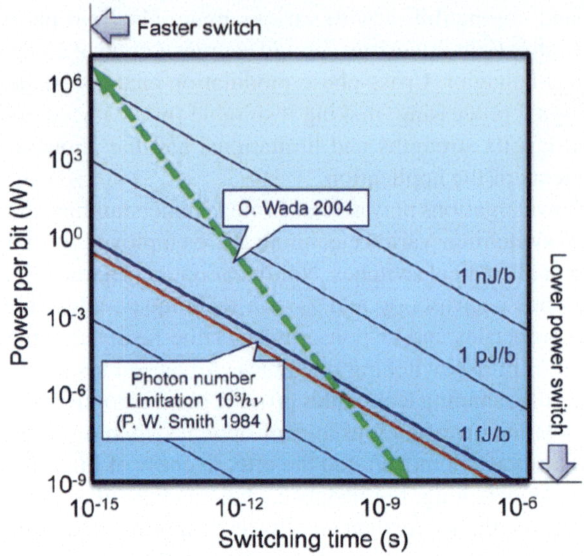

Fig. 8.7 Power per bit as a function of switching time for photonic switches [33]

The results depicted in Fig. 8.8 were recorded by using the experimental configuration Fig. 8.4a where the pump laser beam was fixed at 800 nm while the probe beam wavelength was varied from 480 to 570 nm. When pump and probe laser beams have parallel polarization, the transmission signal was found to be smaller when the probe beam wavelength was increased from 480 to 570 nm Fig. 8.8a. This observation was found to be similar for the case of orthogonal pump and probe beams (Fig. 8.8b). The nondegenerate two-photon absorption coefficients for parallel and orthogonal polarization were estimated by using the concept of crystal orientation and zinc blend symmetry of ZnS single crystals.

The nondegenerate two-photon absorption coefficient is estimated to be about 7.52 cm/GW from Fig. 8.9b, which is consistent with previous reports of literature [26]. It is quite evident from the peak to valley transmission data shown in Fig. 8.9a that there is a large phase change due to the free carrier in comparison to the bound electrons. The value of the nonlinear refraction coefficient is found to be positive. At zero delay, there is a narrow peak for both intensity values, which can be due to the cascading effect of second-order nonlinearities. The carrier recombination time is observed to be 0.6 ns and 0.8 ns for free-carrier absorption and free-carrier refraction, respectively, which can be related to the electrons, holes, and hole-trapping rates in refraction and absorption.

The open-aperture z-scan measurements are carried at input beam energies of 10, 30, and 50 mJ, which corresponds to the incident intensity values of 0.27, 0.82, and 1.37 GW/cm^2, respectively, as shown in Fig. 8.10a. Since the two-photon absorption coefficient values do not vary with the increase in the intensity of the incident beam (Fig. 8.10b). The nonlinear scattering can be estimated by placing a detector at an

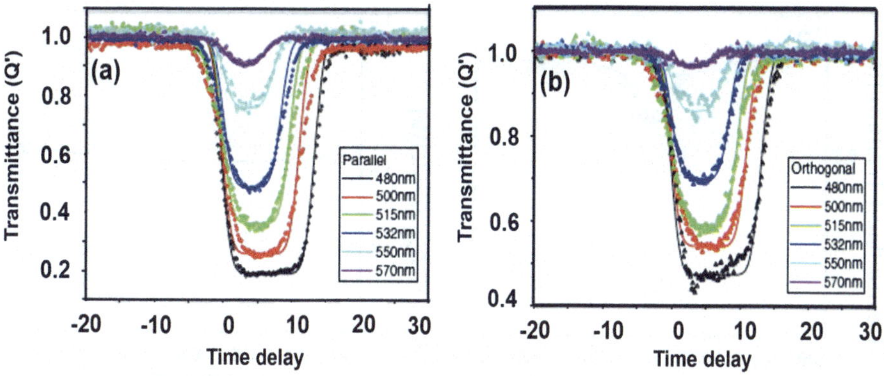

Fig. 8.8 Probe transmission spectra when pump and probe laser beam are in **a** parallel **b** orthogonal polarization mode [5]. These data are recorded by using the experimental configuration shown in Fig. 8.4a

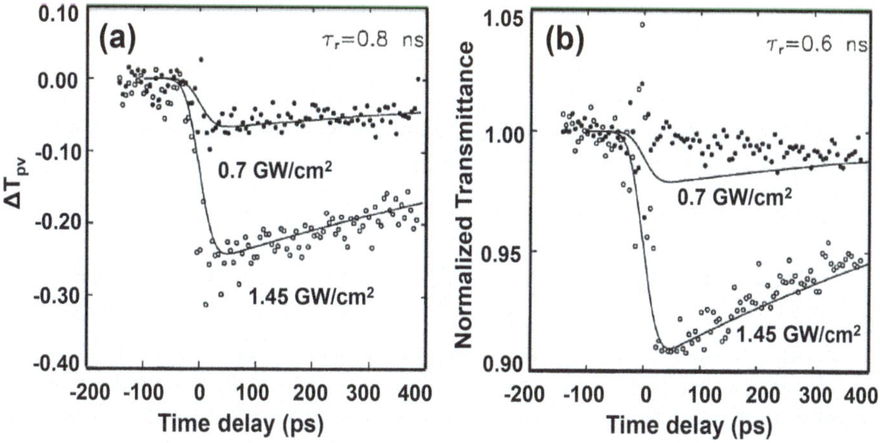

Fig. 8.9 **a** Estimation of nonlinear refraction from peak to valley transmission (ΔTpv). **b** nonlinear absorption from the normalized transmittance as a function of probe beam time delay for ZnS [32]

angle of 45^0 with respect to the incident beam while recording the scattering signal by translating the sample along the propagation direction of the beam. There is almost no change in the signal, which indicates that the nonlinear scattering played hardly a role in the nonlinear absorption measurements. In order to obtain the best options for the material showing optical limiting, the threshold value of incident laser beam intensity and reduction in the transmission play a significant role [6, 24, 31]. It is quite evident that the luminescence quantum yield can be significantly increased by a reduction in the particle size of doped ZnS. Nevertheless, it would be interesting to note whether the reduction of the particle size of the ZnS can also increase the optical limiting behavior of the material. Since the transmittance of material is decreased with the increased intensity of the incident radiation, the optical limiting effect can

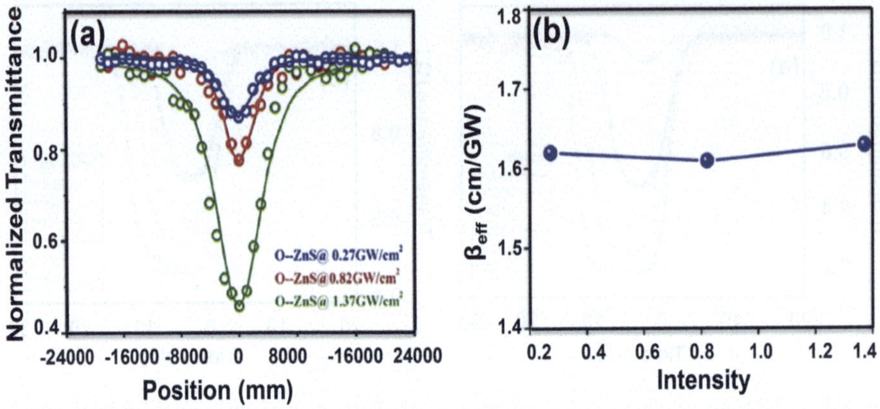

Fig. 8.10 **a** Normalized open-aperture z-scan data of ZnS nanocomposites at different intensities **b** absorption coefficient β_{eff} versus intensity I_0 with units in (GW/cm^2) [9]

be used in various applications where the amount of light entering a system needs to be controlled. This example put forward a significant point in the mind of the scientific community that ZnS can be a candidate for various applications in laser safety devices and protectors from intense radiation.

8.7 Practical Applications of All-Optical Switching Devices

The present section is devoted to discussing the challenges for practical applications of an all-optical switch. The concept of optical transistors and the potential for optical logic circuits to revolutionize information processing are also discussed in this section at length. The question, how to obtain the resolution of the order of attoseconds in an optical switching scheme is also discussed. The attosecond optical switching scheme can be used to fabricate devices with petahertz speeds, which are many orders of magnitude faster than present electronic devices.

To use optical logic for various practical applications, certain conditions must be satisfied, including cascading compatibility, fan-out capabilities, logic-level restoration, input/output isolation, absence of critical biasing, and logic-level independence from loss [25]. In the case of all-optical switching, cascading compatibility is determined by the spatial and temporal profile of the input and output laser pulses. Nevertheless, feedback stages are important for input and output signals in the case of optical transistors [17]. Regarding fan-out capabilities, the output of one stage must drive the inputs of about two or more stages in the next step in the transistors. On the other hand, there must be significant transmission even if the input beam power is low. This is in contrast to existing all-optical switching schemes in the literature, where large irradiance values of the input beam are essential to cause nonlinear changes in the phase of the output beam. The logic level must be restored while the signal propagates in the medium. In the case of all-optical switching-based schemes, the group velocity dispersion of the pulses must be minimized so that we may get a clean signal after transmission.

Another important parameter is input/output isolation, which can be obtained for the transistors; however, it is difficult in optics because many nonlinear processes that occur simultaneously can reduce the final contribution of the signal. In the case of optical signals, the coherent interference of the light beams at different points must be separated, which can justify the absence of critical biasing. Another challenge is logic-level independence from loss in optical switching. It has been observed that the switch-on signals are very clear in the optical switching mechanism; however, the switch-off signals are not as prompt owing to hindrance from various radiative recombination in the semiconductor materials. However, all-optical switching-based devices should have an edge over other switching devices in the future, provided practical implementations are taken into account. It is suggested that the all-optical switching needs further investigation for communication applications.

Most of the findings of researchers in the sub-picosecond all-optical switching domain are in the transmission mode through the target sample, which can attain terabit-rate modulation of the incident signal [19, 29, 30, 34–36]. The main challenge with these types of switches lies in the fact that the intensity of the transmitted signal is quite low, while the switch-off state of the device is not as sharp as that of the switch-on state. One particular advantage is that most of these devices in the sub-picosecond switching time domain have been tested for practical applications at the telecommunication wavelength of 1.55 μm. An example of a recent switch is illustrated in Fig. 8.11a, where the basic principle of an attosecond reflective optical switching device is demonstrated. Reflectivity changes are introduced in the dielectric medium due to the incident pump beam. The pump and probe beams are directed onto the fused silica sample, which is followed by monitoring the reflected probe beam with the help of a photodetector (Fig. 8.11b). The switch-on and switch-off states of the device are shown with attosecond resolution. The attosecond switching as a function of the optical time delay is realized in (Fig. 8.11c). The switching signal is shown in (Fig. 8.11d) when the contribution of the probe beam is subtracted from the overall signal. The switch-on and switch-off states of the attosecond switch are marked with the help of vertical lines using maximum and minimum intensities, respectively. The reflected modulated signal as a function of the optical time delay is shown in Fig. 8.11e. The intensity as a function of wavelength without the pump beam and with the pump beam at zero delay, as well as at an optical delay of 0.9fs are illustrated in Fig. 8.11f. Therefore, reflectivity-based modulation is demonstrated in the fused silica sample under the influence of a highly intense beam, resulting in attosecond optical switching. Furthermore, this information has been used to obtain a data binary encoding system [2, 15, 16]. These original research findings can be used to fabricate devices with petahertz speeds, which are many orders of magnitude faster than present electronic devices.

In the visible region, MnS–ZnS (0.5%) not only shows high photocatalytic efficiency [16] but also the higher values of nonlinear optical parameters, which can be

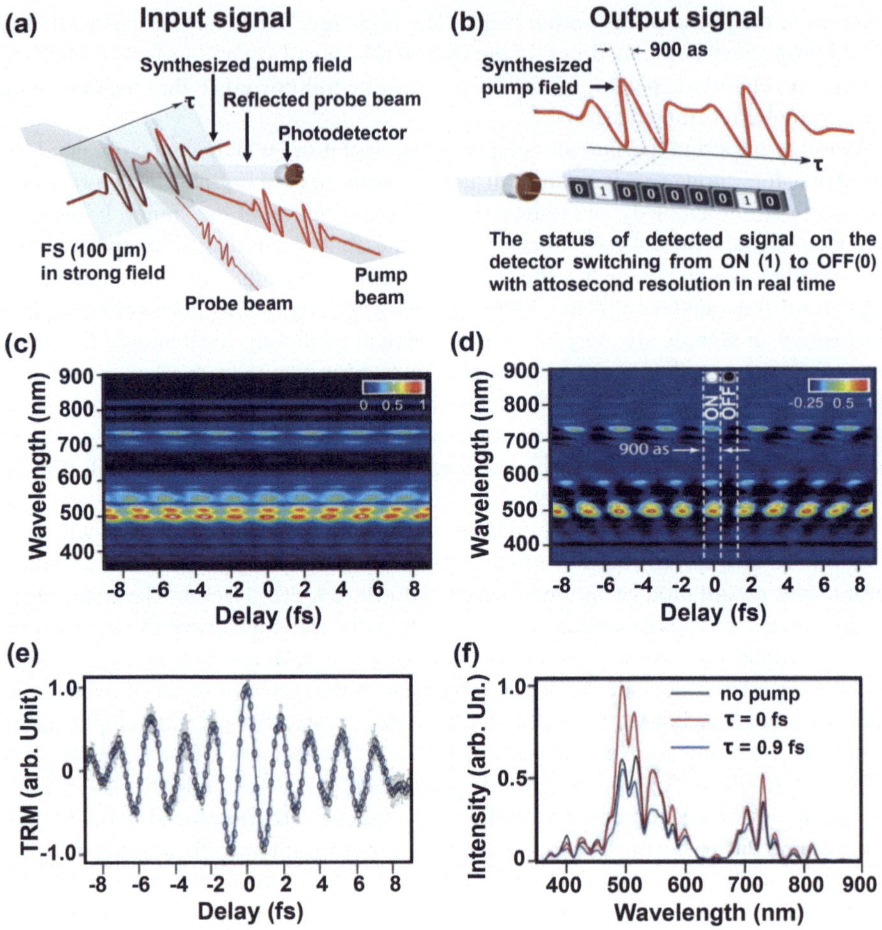

Fig. 8.11 **a** Basic principle of an attosecond reflective optical switching device where pump and probe beams are directed onto a sample of fused silica. **b** The reflected probe beam is monitored with the help of a photodetector. The switch-on and switch-off states of the device are shown with attosecond resolution. **c** attosecond optical switching as a function of the time delay. **d** switching signal when the contribution of the probe beam is subtracted from the overall signal. **e** the reflected modulated signal as a function of the optical time delay. **f** intensity as a function of wavelength without pump beam and with pump beam at zero delay as well as at 0.9 fs [16]

useful to obtain fast optical switching in communication systems [8]. The chalcogenide materials can be used in switching applications where the phase changes are coupled to the cycling ability and non-volatile nature of the material. In addition, low-power consumption and reduced losses are among several other advantages of these materials, which demonstrate their potential for future switching applications [21, 22].

On the other hand, ZnS quantum dots can be utilized in various applications in the field of ultrafast photonics and nonlinear optics. As an example, in the case of the lasers, the InP/ZnSeS/ZnS quantum dot saturable absorber is used to obtain mode-locked laser pulses with a pulse width of 635 fs. The InP/ZnSeS/ZnS quantum dots fabricated in the presence of hydrogen fluoride can be used to reduce defects, which leads to a reduction of the carrier recovery time. The pulse width of the laser is shortened from 635 fs to 450 fs by using InP/ZnSeS/ZnS quantum dots [23].

8.8 Summary

The present chapter has provided a comprehensive overview of the mechanisms and experimental configurations of all-optical switching. High figure-of-merit materials play a pivotal role in enabling efficient and reliable switching operations. The different mechanisms of all-optical switching offer unique advantages, and their choice depends on the specific requirements of the application. Experimental techniques play a vital role in studying and understanding all-optical switching phenomena, providing insights into material properties and performance. Recent experimental studies have demonstrated the potential of high FOM materials in achieving efficient all-optical switching, paving the way for the development of faster, more reliable, and energy-efficient communication systems.

References

1. H.-S. Albrecht, P. Heist, J. Kleinschmidt, D.V. Lap, Ultrafast beam-deflection method and its application for measuring the transient refractive index of materials. Appl. Phys. B **57**, 193–197 (1993)
2. H. Alqattan, D. Hui, M. Sennary, M.T. Hassan, Attosecond electronic delay response in dielectric materials. Faraday Discus. **237**, 317–326 (2022)
3. R.W. Boyd, A.L. Gaeta, E. Giese, Nonlinear optics, in *Springer Handbook of Atomic, Molecular, and Optical Physics* (Springer, 2008), pp. 1097–1110
4. P.N. Butcher, D. Cotter, *The Elements of Nonlinear Optics* (Cambridge University Press, 1990)
5. S. Chen, M.-L. Zheng, X.-Z. Dong, Z.-S. Zhao, X.-M. Duan, Nondegenerate two-photon absorption in a zinc blende-type zns single crystal using the femtosecond pump-probe technique. JOSA B **30**(12), 3117–3122 (2013)
6. D.N. Christodoulides, I.C. Khoo, G.J. Salamo, G.I. Stegeman, E.W. Van Stryland, Nonlinear refraction and absorption: mechanisms and magnitudes. Adv. Opt. Photon. **2**(1), 60–200 (2010)
7. C. Corrado, Y. Jiang, F. Oba, M. Kozina, F. Bridges, J.Z. Zhang, Synthesis, structural, and optical properties of stable zns: Cu, cl nanocrystals. J. Phys. Chem. A **113**(16), 3830–3839 (2009)
8. D.V. Dake, N.D. Raskar, V.A. Mane, R.B. Sonpir, H.A. Khawal, U. Deshpande, E. Stathatos, B.N. Dole, Photocatalytic performance of graphene-based cr-substituted β zns nanocomposites. Appl. Phys. A **128**(4), 276 (2022)

9. I. Davydenko, S. Barlow, R. Sharma, S. Benis, J. Simon, T.G. Allen, M.W. Cooper, V. Khrustalev, E.V. Jucov, R. Castaneda, et al. Facile incorporation of pd (pph3) 2hal substituents into polymethines, merocyanines, and perylene diimides as a means of suppressing intermolecular interactions. J. Am. Chem. Soc. **138**(32), 10112–10115 (2016)
10. R. DeSalvo, A.A. Said, D.J. Hagan, E.W. Van Stryland, M. Sheik-Bahae, Infrared to ultraviolet measurements of two-photon absorption and n/sub 2/in wide bandgap solids. IEEE J. Quantum Electron. **32**(8), 1324–1333 (1996)
11. M.R. Ferdinandus, H. Hu, M. Reichert, D.J. Hagan, E.W. Van Stryland, Beam deflection measurement of time and polarization resolved ultrafast nonlinear refraction. Opt. Lett. **38**(18), 3518–3521 (2013)
12. J.M. Hales, J. Matichak, S. Barlow, S. Ohira, K. Yesudas, J.-L. Brédas, J.W. Perry, S.R. Marder, Design of polymethine dyes with large third-order optical nonlinearities and loss figures of merit. Science **327**(5972), 1485–1488 (2010)
13. R.A. Hann, D. Bloor, *Organic Materials for Non-linear Optics II*, vol. 2 (Royal Society of Chemistry, 1991)
14. S.A. Haque, J. Nelson, Toward organic all-optical switching. Science **327**(5972), 1466–1467 (2010)
15. D. Hui, H. Alqattan, S. Yamada, V. Pervak, K. Yabana, M.Th. Hassan, Attosecond electron motion control in dielectric. Nat. Photon. **16**(1), 33–37 (2022)
16. D. Hui, H. Alqattan, S. Zhang, V. Pervak, E. Chowdhury, M.Th. Hassan, Ultrafast optical switching and data encoding on synthesized light fields. Sci. Adv. **9**(8), eadf1015 (2023)
17. J. Hwang, M. Pototschnig, R. Lettow, G. Zumofen, A. Renn, S. Götzinger, V. Sandoghdar, A single-molecule optical transistor. Nature **460**(7251), 76–80 (2009)
18. C.-Y. Jin, O. Wada, Photonic switching devices based on semiconductor nano-structures. J. Phys. D: Appl. Phys. **47**(13), 133001 (2014)
19. X. Jiao, Y. Peng, S. Qian, L. Jiang, Microstructured all-optical switching based on two-dimensional material. Coatings **13**(5), 876 (2023)
20. S. Kim, R. Sharma, B. Kim, H.-S. Yang, K. Kyhm, Modal gain enhancement by cylindrical waveguide and gain saturation in cdse nanocrystal quantum dots. J. Phys. D: Appl. Phys. **42**(9), 095403 (2009)
21. J. Lapointe, C. Coia, A. Dupont, R. Vallée, Passive broadband faraday isolator for hybrid integration to photonic circuits without lens and external magnet. Nat. Photon. 1–10 (2025)
22. C. Li, R. Pan, G. Changzhi, H. Guo, J. Li, Reconfigurable micro/nano-optical devices based on phase transitions: From materials, mechanisms to applications. Adv. Sci. **11**(20), 2306344 (2024)
23. Y. Lou, P. He, B. Ge, X. Duan, H. Lei, X. Zhang, L. Tao, X. Zhang, Inp/znses/zns core-shell quantum dots as novel saturable absorbers in mode-locked fiber lasers. Adv. Opt. Mater. **11**(3), 2201939 (2023)
24. S.R. Marder, Materials for third-order nonlinear optics. MRS Bull. **41**(1), 53–62 (2016)
25. D. Miller, Are optical transistors the logical next step? Nat. Photon. **4**(1), 3–5 (2010)
26. R.A. Negres, J.M. Hales, A. Kobyakov, D.J. Hagan, E.W. Van Stryland, Two-photon spectroscopy and analysis with a white-light continuum probe. Opt. Lett. **27**(4), 270–272 (2002)
27. D.J. Norris, A.L. Efros, S.C. Erwin, Doped nanocrystals. Science **319**(5871), 1776–1779 (2008)
28. B.V. Olson, M.P. Gehlsen, T.F. Boggess, Nondegenerate two-photon absorption in gasb. Opt. Commun. **304**, 54–57 (2013)
29. S. Saha, B.T. Diroll, M.G. Ozlu, S.N. Chowdhury, S. Peana, Z. Kudyshev, R.D. Schaller, Z. Jacob, V.M. Shalaev, A.V. Kildishev et al., Engineering the temporal dynamics of all-optical switching with fast and slow materials. Nat. Commun. **14**(1), 5877 (2023)
30. R. Sharma, J. Kim, H.-S. Yang, K. Kyhm, Sub-picosecond all-optical switching of exciton spin polarization in nanocrystal quantum dots toward terabit communication. J. Korean Phys. Soc. **55**(1), 259–262 (2009)
31. M. Sheik-Bahae, A.A. Said, T.-H. Wei, D.J. Hagan, E.W. Van Stryland, Sensitive measurement of optical nonlinearities using a single beam. IEEE J. Quantum Electron. **26**(4), 760–769 (1990)

32. M. Sheik-Bahae, J. Wang, E.W. Van Stryland, Nondegenerate optical Kerr effect in semiconductors. IEEE J. Quantum Electron. **30**(2), 249–255 (1994)
33. P.W. Smith, Applications of all-optical switching and logic, in *Philosophical Transactions of the Royal Society of London*. Series A, Mathematical and Physical Sciences, vol. 313(1525), pp. 349–355 (1984)
34. Y. Sun, A. Larin, A. Mozharov, E. Ageev, O. Pashina, F. Komissarenko, I. Mukhin, M. Petrov, S. Makarov, P. Belov, et al., All-optical generation of static electric field in a single metal-semiconductor nanoantenna. Light: Sci. Appl. **12**(1), 237 (2023)
35. L. Wenting Wang, Q.Z. Huang, X. Yuan, M. Ali, Q. Meng, W. Sun, Ultrafast third-order nonlinear optical response of ag modified znse/zns core-shell structure quantum dots composite films. Optik **252**, 168537 (2022)
36. P. Zhang, T.-F. Chung, Q. Li, S. Wang, Q. Wang, W.L.B. Huey, S. Yang, J.E. Goldberger, J. Yao, X. Zhang, All-optical switching of magnetization in atomically thin cri3. Nat. Mater. **21**(12), 1373–1378 (2022)
37. P. Zhao, M. Reichert, S. Benis, D.J. Hagan, E.W. Van Stryland, Temporal and polarization dependence of the nonlinear optical response of solvents. Optica **5**(5), 583–594 (2018)
38. J. Zheng, Z. Yang, C. Si, Z. Liang, X. Chen, R. Cao, Z. Guo, K. Wang, Y. Zhang, J. Ji et al., Black phosphorus based all-optical-signal-processing: toward high performances and enhanced stability. ACS Photon. **4**(6), 1466–1476 (2017)

Chapter 9
Challenges and Future Directions in All-Optical Switching

9.1 Wavelength Swapping and Nonlinear Absorption

The pump-probe measurements of the ZnS sample are conducted using pump-probe beam wavelengths of 800 nm and 400 nm, respectively. The question of whether swapping the wavelengths of the pump and probe beams in nondegenerate pump-probe measurements can provide additional information about the relaxation time of hot electrons in the conduction band will be discussed. Another question we explore is whether the swapping exercise can be utilized to determine the characteristics of ZnS for application in optical switching devices.

In the first step, the pump and probe beams were kept in a similar polarization state, with wavelength values of 800 nm and 400 nm, respectively. This situation is highlighted in Fig. 9.1a and can be considered as an example of the nondegenerate two-photon absorption process. It is easy to see that there is no transition from subband 1 to subband 2, as three-photon absorption is neglected. On the other hand, one can observe that the linewidth of the pump-probe transmission signal (Fig. 9.1b) is significantly presented in the inset of Fig. 9.1b This can be explained on the basis of group velocity dispersion, where the 800 nm laser beam travels faster than the 400 nm laser beam due to the difference in the values of refractive indices. Therefore, the pump and probe laser beams cannot be at zero delay, as predicted by the cross-correlation measurements. One can estimate the nondegenerate two-photon absorption coefficient value using the following relation 9.1:

$$\Gamma = L(n/\,n_{pr\,pu})I_{pu}{}^{0}\beta \tag{9.1}$$

where, L, n_{pu}, n_{pr}, $I^0{}_{pu}$ are the sample thickness, linear refractive index of pump beam, linear refractive index of probe beam, and peak irradiance of pump beam, respectively. In order to record the results of Fig. 9.2a the probe beam intensity is

R. Sharma, *Nonlinear Optical Materials for All-Optical Switching Applications*, Progress in Optical Science and Photonics 33,
https://doi.org/10.1007/978-981-96-5921-0_9

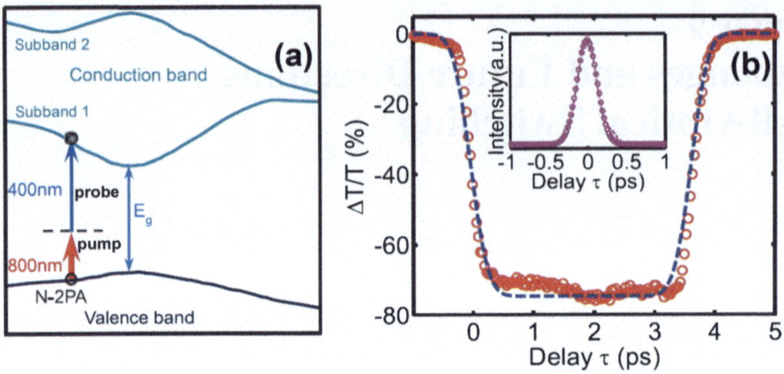

Fig. 9.1 **a** The band-structure diagram of nondegenerate two-photon absorption in ZnS, where pump and probe laser beams are kept at a wavelength of 800 and 400 nm, respectively. **b** Percentage change in transmission as a function of time delay when probe beam is kept at a wavelength of 400 nm. The inset shows the temporal profile of the fundamental laser beam at a wavelength of 800 nm. The full width at half maximum of the femtosecond laser beam is found to be 200 fs [18]

fixed while the pump intensity is varied from 0.22 GW/cm^2 to 2.24 GW/cm^2. When the time delay is at 0 ps, a sharp decrease in the transmission is seen; however, the curves remain flat from 0.5 to 3 ps. After 3.5 ps, the transmission increases to the original level without showing any tail effect. One can easily recall that the value of the nondegenerate two-photon coefficient does not vary despite the increase in the pump beam intensity, as illustrated by Fig. 9.2b. The free-carrier absorption, degenerate two-photon absorption, and nondegenerate two-photon absorption processes are schematically illustrated in Fig. 9.2c. When pump and probe beams are respectively kept at a wavelength of 400 and 800 nm, the energy bandgap of ZnS is covered by the nondegenerate two-photon absorption. On the other hand, when the energy bandgap of ZnS is covered with two photons, each of wavelength 400 nm, the process is called the degenerate two-photon absorption. The free-carrier absorption is represented by the excitation of electrons in the conduction band from subband 1 to subband 2. In order to find the answer to the question "how the hot electrons in the conduction are related to the minimum energy state," swapping of the pump and probe beam wavelengths can be helpful, which is illustrated in the transmission results of Fig. 9.2d. The inset of Fig. 9.2b is a magnified version of the nonlinear transmission around zero delay. Since the transmission signal is reduced by about 9% around zero delay, it emphasizes that not only the nondegenerate two-photon absorption but also the degenerate two-photon absorption process contributes to the overall signal. There is a long tail associated with the time delay of 4 ps, which is related to the free-carrier lifetime of hot electrons in subband 1. Recalling the fact that the free-carrier lifetime is larger than the pulse width (200 fs) of the laser beam, the lifetime can be obtained by an exponential to the recovery time of switching to the initial state. The lifetime measurements include a sum of radiative and non-radiative transitions in the

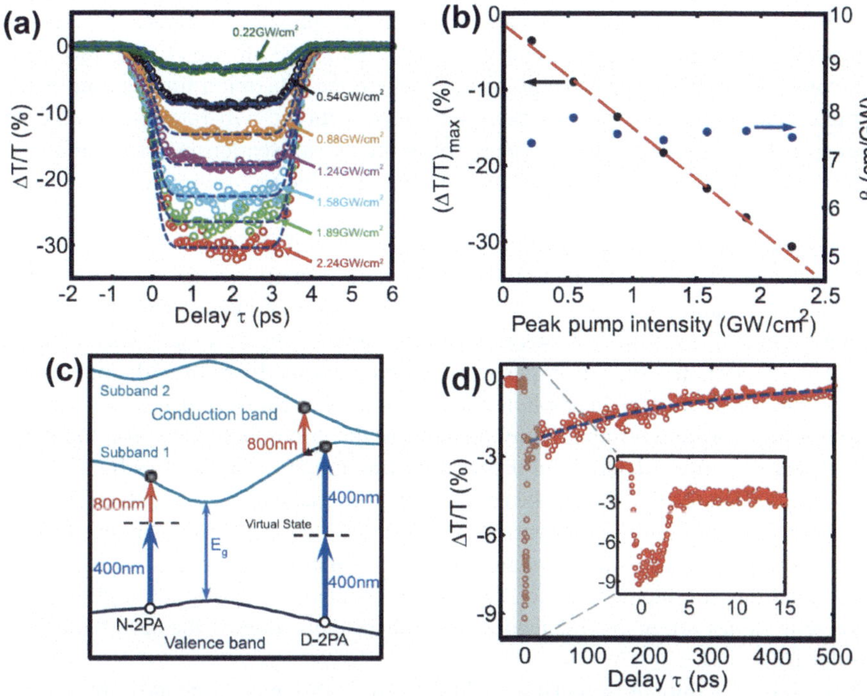

Fig. 9.2 **a** Transmittance change as a function of time delay when the pump and probe beam are kept at wavelengths of 800 and 400 nm, respectively. The pump beam intensity is varied from 0.22 to 2.24 GW/cm^2. **b** Maximum transmission and nondegenerate two-photon absorption coefficients for various intensities of pump beam. **c** Energy subband structure diagram depicting free-carrier absorption, degenerate two-photon absorption, and nondegenerate two-photon absorption processes. **d** The relative transmission as a function of time delay when the pump and probe beams are respectively, kept at 400 and 800 nm wavelength. The inset shows the magnified view of the main level near zero delay [18]

materials, which can be deconvoluted to extract the various components. Multiexponential decay curves indicate more than one energy state, which is responsible for the decay process. On the other hand, single exponential decay means that only one state is responsible for decay measurements. Nevertheless, the relaxation processes at a scale of ns and even hundreds of ps can hinder the switch-off state which resulting in an incomplete switch-off state. In such cases, the switch-on state is sharp enough; however, the switch-off state is not complete, which results in reduced contrast of the switch.

Further studies are required to investigate the effect of the electron-phonon interaction in ZnS. On the other hand, the experimental result in silicon semiconductor shows that the photo-excitation leads to a decrease in the relaxation rates. This can reduce the carrier temperature through phonon scattering. In order to get the relaxation properties of the hot electrons at relatively higher energy states, the wavelengths of the pump and probe beams are swapped. When the optical delay between the pump and probe is kept around 4 ps, the absorption of the probe photons by hot electrons dominates. We believe that the materials, in which the relaxation time is dominated by the electron-electron and electron-phonon transitions, will be relatively better candidates for switching applications because the carrier scattering occurs at a sub-picosecond time scale. On the other hand, if the radiative recombination dominates, then it results in the longer lifetime values (ns) of material, which produces a long tail, and the switch-off state is prolonged over the period of time.

Next, we discuss the question of whether the nonlinear absorption is affected if the ZnS nanosheet samples are annealed. The situation is highlighted in Fig. 9.3a, where transmission as a function of Z-position is illustrated for the ZnS nanosheet samples annealed at 400 °C for 2h. The experimental data are fitted by both of the processes, the two-photon as well as three-photon absorption. Nevertheless, it is found that the two-photon absorption fitting is better than that of the three-photon absorption fitting. This aspect is also validated from the straight line graph of ln(T_{OA}) vs ln(I). There are two important results that can be drawn from the z-scan measurements of the ZnS nanosheet. The two-photon absorption is found to be about 10 times higher in ZnS nanocrystal quantum dots as compared to the ZnS nanosheet sample, which is due to the quantum confinement effect. On the other hand, the two-photon absorption coefficient values are also enhanced as a result of annealing of the ZnS nanosheet samples [6, 14, 18].

The nonlinear optical properties of Mn-doped ZnS nanocomposites are investigated to explore the influence of Mn in the host material [4]. Third-order nonlinear optical susceptibility and nonlinear refractive index values are shown for pure and doped samples of ZnS in Fig. 9.4. It is quite evident that the doped samples have shown smaller values of the nonlinear refractive index as compared to the undoped

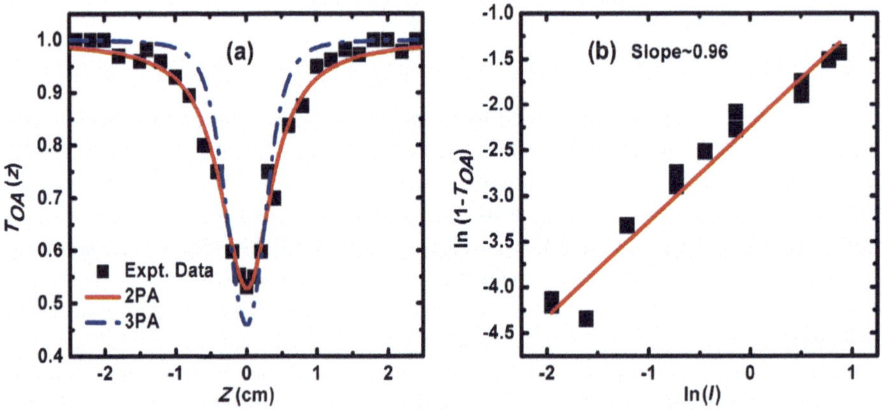

Fig. 9.3 **a** Transmission versus Z-position of ZnS–ZnO nanocomposite sample in an open-aperture z-scan measurements when intensity beam is kept at 2.41 GW/cm^2. The symbols represent the experimental data while the two and three-photon absorption fittings are shown with the help of solid and dashed lines, respectively **b** time of annealing (T_{OA}) as a function of the intensity [8]

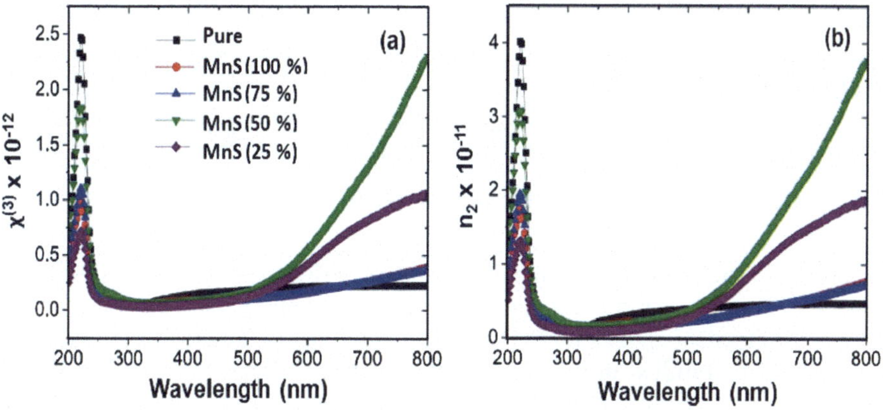

Fig. 9.4 **a** Third-order nonlinear optical susceptibility and **b** nonlinear refractive index for pure and MnS/ZnS nanocomposites [4]

samples of ZnS, which is an opposite trend in the case of the luminescent properties of the host materials, where luminescent properties are enhanced as a result of doping.

Since the transmittance of material is decreased with the increased intensity of the incident radiation, the optical limiting effect can be used in various applications where the amount of light entering a system needs to be controlled. In this regard, the two-photon absorption process in ZnS can be useful to extract information on whether the material can be a suitable candidate for detector and optical limiting applications. The higher the value of the nonlinear absorption coefficient, the greater the possibility of a material being used in detector and optical limiting applications. Two important points can be noticed from the nonlinear absorption coefficient measurements: (1) If a material has a high value of nonlinear absorption coefficient, then it is not suitable for all-optical switching applications because the figure-of-merit is

Fig. 9.5 Estimation of figure-of-merit [7]

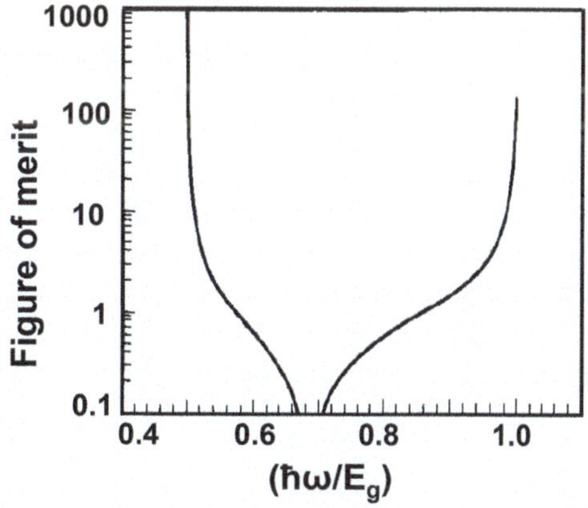

inversely proportional to the value of nonlinear absorption coefficient. (2) Despite a low figure-of-merit, the material still remains useful for detector and optical limiting applications.

9.2 Challenges in All-Optical Switching Devices

According to the scientific commentary of Miller [13], there would be significant challenges for the scientific community to use the femtosecond laser-based AOS devices for practical applications. The main conditions for the practical logic operation should be fulfilled in the near future, which are discussed here. The spatial and temporal profile of the input and the output beams can be maintained by the use of modern technology such as solitons [11]. The amplification process would be sufficient to drive more than two stages. The beam and signal qualities must be maintained during the propagation of the signal. However, the input isolation, the minimum losses, and preservation of the relative phases over the long distances of the coherent interference of the light beams may pose significant challenges for the researchers. The input power and the contrast of all-optical logic gates made up of networks of the optical cavities with second and third-order optical nonlinearities were investigated, where second-order optical nonlinearity-based photonic circuits were found to be an order of magnitude energy efficient than the third-order nonlinearity-based photonic circuits at the quality factor value of 10^4. Previously, the fiber amplifiers were doped with certain impurities to make them readily useful for silicon-organic all-optical switches.

Nevertheless, the fabrication was tedious, which also makes the expensive device. The Mach-Zehnder configuration based on the nonlinear refractive index values will be useful to directly investigate the output signal of the all-optical switch. A high nonlinearity of the organic materials coupled with the silicon integrated circuit can be used as a best combination for the optical switching applications [1, 2, 9, 10, 12]. There are multiple advantages of the organic materials such as large third-order nonlinear optical response, user friendly molecular design, stable nature corresponding to high irradiance etc. Two signals A and B are injected into the switch, which directs the output signal to the terminal C. However, the signal from terminal C is switched to terminal D on the application of the applied voltage across the device. This fact is illustrated by Fig. 9.6a, which originates from the change in the refractive index value on the application of the bias voltage. On the other hand, a similar effect is noticed in Fig. 9.6b; however, the nonlinear refractive index changes are induced by the optical signals. The switching time of the all-optical device can be of the order of fs. Nevertheless, in the case of the electro-optic modulators, the switching time can be of the order of ns because the electro-optic response of the material is slower than the all-optical response.

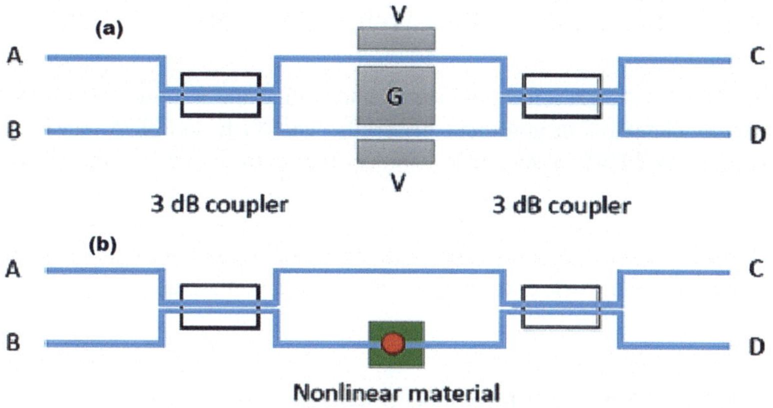

Fig. 9.6 **a** Electro-optic modulator with Mach-Zehnder configuration and **b** organic molecule based Mach-Zehnder configuration. Both are forms of the switches

9.3 Possible Solutions to Overcome Challenges

The AOS is an important component of the high-speed optical communication network where the signal beam is modulated by another beam, the so-called control beam, by exploiting the nonlinear properties of the material. Recalling the fact that the switching depends on the materials that have not only the high NLR but also have low NLA values. The optical switching devices based on different materials were used, such as semiconductor, organic molecules, chalcogenide glasses, nanoparticles, etc. All types of optical switches work on the principle of change of the refractive index of the NLO materials due to the nonlinear susceptibility, although there are different mechanisms responsible for changing the NLR of the material.

Next, the ultrafast AOS in high FOM YZ-V-69 organic dye is realized by using the interference fringe pattern of the MZI. This is realized that at 6, 9 and 12 GW/cm^2 control laser beam powers are not suitable for AOS operation in YZ-V-69 organic molecule, because, the δ_{NLR} is low which is not sufficient to obtain π-phase shifts [15]. At 21 GW/cm^2, the δ_{NLR} is high, which results in the phase shift; however, the central maxima is not completely converted into minima. This is confirmed by the comparison of 18 and 21 GW/cm^2 powers where the phase modulation symmetry is maintained.

This is to conclude that the 18 GW/cm^2 is a suitable control beam irradiance to obtain AOS operation in the investigated molecule rather than the other lower values of irradiance. The phase change about π introduced by the control pulses with irradiance value of 18 GW/cm^2, which is attributed to a high figure-of-merit value of YZ-V-69 organic dye.

9.4 Future Directions of Research in All-Optical Switching

The objectives of the present book are to understand the nonlinear refraction and absorption mechanisms in nonlinear materials. The NLR and NLA can be further used to obtain the FOM value, which can decide the suitability of material for AOS applications. The off-resonant (degenerate), free space optics, and MZI-based all-optical switch configuration have been developed. The next challenge is to check the material for realistic application using silicon-organic hybrid waveguide and to look for the nondegenerate NLR enhancement in the organic molecule [14].

9.4.1 AOS of YZ-V-69 Using Silicon Organic Hybrid Waveguide

If YZ-V-69 organic material is incorporated in a slot waveguide geometry, then it can be a suitable candidate for all-optical switching applications. The FOM value of about ≥ 1 is reasonable for the AOS applications; however, we have seen that the YZ-V-69 materials have an FOM value greater than ≥ 400. We believe that the high FOM materials can be selected for the AOS switching applications by using a Mach-Zehnder interferometry configuration, which can be further extended to the telecommunication spectral window by incorporating the high FOM material in the slot waveguide geometry.

9.4.2 Nondegenerate NLR Enhancement in Material

The END-2PA coefficient values in the ZnSe sample are enhanced about 270 times compared to the D-2PA and ND-2PA coefficient values for the particular selection of the wavelengths. This type of selection of wavelengths is also proposed for obtaining high NLR values of the NLO materials [14]. The nondegenerate nonlinear refraction measured using the two-color Z-scan technique [16]. They have measured only for CS2 (n_2) and ZnSe (α_2) at single value of probe wavelength. The color two Z-scan has certain limitations, such as this technique does not provide temporal dynamics of the material and fails to separate out the different kinds of nonlinear mechanisms which are responsible for nonlinear optics. The value of ultrafast nonlinearity obtained by Z-scan is the average of both the slow as well fast effects of the material. Further, this issue is resolved by the beam-deflection technique [5] where the pump and probe beams are utilized to extract the transient response of the material. This is also a good project to characterize organic molecules with the beam-deflection technique to obtain the temporal dynamics.

9.5 Strategies to Enhance Figure-of-Merit

In the photonic devices, the higher power consumption and thermal losses hinder most of the practical applications. All-dielectric meta-materials were investigated for the switching applications because these photonic devices consume less power, and the thermal losses are minimal. In this regard, ZnS/SiO_2 was considered as one of the materials that can be embedded in photonic devices [17]. These all-dielectric switches can attain picosecond switching speed, which can be further increased to femtosecond switching speed with smart selection of the incident beam wavelength and manipulation of the material by doping [3]. The linear absorption can be kept low in case of the nonresonant optical nonlinearity, therefore, the transmission of light is increased. In case of the nonresonant optical nonlinearity, nonlinear refractive index, n_2, and the nonlinear absorption coefficient, α_2, values can be related to the Kramers-Kronig relation for a broad range of solids, as illustrated in Fig. 9.5. It has to be emphasized that the response time of the semiconductor to the incident field is of the order of 10^{-15} s, which can be considered as an instantaneous response.

The schematic evolution of the idea of switching in the present investigations is depicted in Fig. 9.7, where pristine ZnS shows very weak luminescence, which can be considered to originate from defects, band-edge energy states, and excitonic energy levels. The weak emission peak is centered around 420 nm and exhibits luminescence lifetime values of the order of microseconds. This situation is highlighted in Fig. 9.7a. However, luminescence quantum yield is significantly increased when using Mn dopant, owing to the introduction of $^4T_1$ and $^6A_1$ energy states in the host (Fig. 9.7b). The strong luminescence peak is centered around 585 nm with luminescence lifetime values on the order of milliseconds. Thus, the phosphor nature of ZnS is significantly enhanced by the incorporation of dopants, which can be utilized in various optoelectronic applications. Nevertheless, the situation is drastically altered when the irradiance value of the incident beam on ZnS is high enough to cause nonlinear optical effects. Specifically, when the energy bandgap of ZnS is covered with two photons of the same wavelength, the process is called degenerate two-photon absorption, which is used for various optical limiting applications, as illustrated by Fig. 9.7c. Recalling the fact that two-photon absorption contributes to material losses, it is undesirable in the case of switching applications. In the search for schemes to achieve a high figure-of-merit, losses are reduced by using the wavelength of incident radiation where one-photon and two-photon absorption processes are avoided, as shown in Fig. 9.7d. The idea can be extended to a broad range of materials by using the scheme shown in Fig. 9.7e, so that the nonlinear refractive index coefficient values are increased. The all-optical switching mechanism in nanocrystal quantum dots is highlighted in Fig. 9.7f, where the spin polarization concept is used.

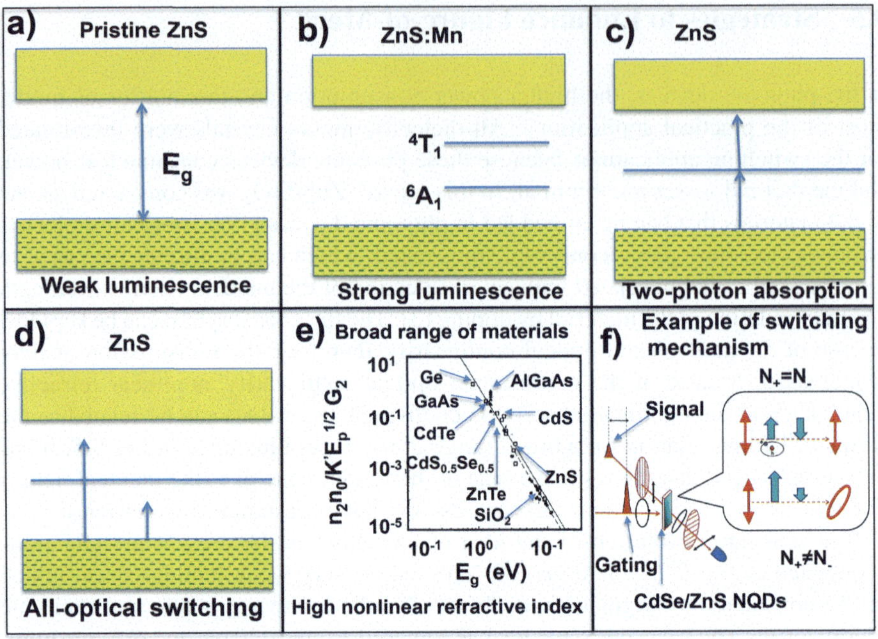

Fig. 9.7 Schematic evolution of the idea of switching in the present investigations: **a** Pristine ZnS has a very low luminescence quantum yield, which can be considered to originate from defects, band-edge energy states, and excitonic energy levels. **b** Luminescence quantum yield is significantly increased when using Mn dopant, owing to the introduction of $^4T_1$ and $^6A_1$ energy states in the host. **c** When the energy bandgap of ZnS is covered with two photons of the same wavelength, the process is called degenerate two-photon absorption, which is used for various optical limiting applications. **d** In the search for schemes to achieve a high figure-of-merit, losses are reduced by using the wavelength of incident radiation where one-photon and two-photon absorption processes are avoided. **f** The idea can be extended to a broad range of materials, increasing the nonlinear refractive index coefficient values [7]. **g** The all-optical switching mechanism in nanocrystal quantum dots, where the spin polarization concept is used

9.6 Summary

The present chapter highlighted various challenges in the development and implementation of all-optical switching in high figure-of-merit material, which is followed by exploring possible solutions to overcome these challenges. Also, future directions of research in all-optical switching in high figure-of-merit materials are discussed.

References

1. E. Adibnia, M. Ghadrdan, M.A. Mansouri-Birjandi, Nanophotonic structure inverse design for switching application using deep learning. Sci. Rep. **14**(1), 21094 (2024)
2. E. Adibnia, M.A. Mansouri-Birjandi, M. Ghadrdan, P. Jafari, A deep learning method for empirical spectral prediction and inverse design of all-optical nonlinear plasmonic ring resonator switches. Sci. Rep. **14**(1), 5787 (2024)
3. A. Chauhan, R. Sharma, M. Singh, R. Sharma, Optical non-linearities and applications of zns phosphors. Adv. Opt. Technol. **13**, 1390474 (2024)
4. A.M. El-Naggar, Z.K. Heiba, M. Bakr Mohamed, A.M. Kamal, G. Lakshminarayana, O.H. Abd-Elkader, Effect of mns/zns nanocomposite on the structural, linear and nonlinear optical properties of pva/cmc blended polymer. Opt. Mater. **128**, 112379 (2022)
5. M.R. Ferdinandus, H. Hu, M. Reichert, D.J. Hagan, E.W. Van Stryland, Beam deflection measurement of time and polarization resolved ultrafast nonlinear refraction. Opt. Lett. **38**(18), 3518–3521 (2013)
6. D.A. Fishman, C.M. Cirloganu, S. Webster, L.A. Padilha, M. Monroe, D.J. Hagan, E.W. van Stryland, Sensitive mid-infrared detection in wide-bandgap semiconductors using extreme non-degenerate two-photon absorption. Nat. Photon. **5**(9), 561–565 (2011)
7. C.N. Ironside, Ultra-fast all-optical switching. Contemp. Phys. **34**(1), 1–18 (1993)
8. A.K. Kole, P. Kumbhakar, U. Chatterjee, Observations on nonlinear optical properties of zns nanosheet, zns–zno composite nanosheet and porous zno nanostructures dispersed in aqueous medium. Chem. Phys. Lett. **591**, 93–98 (2014)
9. C. Koos, P. Vorreau, T. Vallaitis, P. Dumon, W. Bogaerts, R. Baets, B. Esembeson, I. Biaggio, T. Michinobu, F. Diederich et al., All-optical high-speed signal processing with silicon-organic hybrid slot waveguides. Nat. Photon. **3**(4), 216–219 (2009)
10. R. Kumari, A. Yadav, B. Lahiri, Photonic crystal all-optical switches. *Optical Switching: Device Technology and Applications in Networks*, pp. 147–164 (2022)
11. F. Kurtz, C. Ropers, G. Herink, Resonant excitation and all-optical switching of femtosecond soliton molecules. Nat. Photon. **14**(1), 9–13 (2020)
12. M. Maiuri, A. Schirato, G. Cerullo, G. Della Valle, Ultrafast all-optical metasurfaces: challenges and new frontiers. ACS Photon. **11**(8), 2888–2905 (2024)
13. D. Miller, Are optical transistors the logical next step? Nat. Photon. **4**(1), 3–5 (2010)
14. H.S. Pattanaik, M. Reichert, J.B. Khurgin, D.J. Hagan, E.W. van Stryland, Enhancement of two-photon absorption in quantum wells for extremely nondegenerate photon pairs. IEEE J. Quantum Electron. **52**(3), 1–14 (2016)
15. R. Sharma, M. Singh, H. Kaur, Recent advances in high-figure-of-merit semiconductor and organic materials for all-optical switching applications. J. Mater. Sci. **56**(4), 2838–2855 (2021)
16. M. Sheik-Bahae, A.A. Said, T.-H. Wei, D.J. Hagan, E.W. van Stryland, Sensitive measurement of optical nonlinearities using a single beam. IEEE J. Quantum Electron. **26**(4), 760–769 (1990)
17. L. Wenting Wang, Q.Z. Huang, X. Yuan, M. Ali, Q. Meng, W. Sun, Ultrafast third-order nonlinear optical response of ag modified znse/zns core-shell structure quantum dots composite films. Optik **252**, 168537 (2022)
18. Y.-E. Wu, M. Ren, Z. Wang, W. Li, Q. Wu, S. Yi, X. Zhang, J. Xu, Optical nonlinear dynamics in ZnS from femtosecond laser pulses. AIP Adv. **4**(5) (2014)

Chapter 10
Coherent Anti-Stokes Raman Scattering (CARS) Microscopy: Instrumentation, Theory, and Applications

10.1 Introduction to CARS Microscopy

Coherent Anti-Stokes Raman Scattering (CARS) microscopy is a transformative approach to vibrational imaging, which offers high sensitivity, rapid imaging speeds, and three-dimensional spatial resolution. This technology has enabled new possibilities in chemical and biological imaging, facilitating the visualization of molecular dynamics and distributions with minimal sample perturbation. By combining advanced laser optics and nonlinear optical processes, CARS microscopy has evolved into a powerful tool for studying complex systems, such as living cells, with exceptional chemical specificity and spatial precision. CARS microscopy relies on a four-wave mixing process where incident laser fields generate a coherent anti-Stokes signal [7, 8, 12]. This signal arises from the interaction of pump, Stokes, and probe fields within a sample, resonating with molecular vibrations to produce a characteristic spectral signature. Compared to fluorescence microscopy, which requires labeling, or spontaneous Raman microscopy, which suffers from low sensitivity, CARS microscopy provides noninvasive, label-free imaging with superior speed and resolution [2, 5, 10, 11, 13].

10.2 Theoretical Foundations and Mechanisms

The core of CARS microscopy lies in its nonlinear optical properties. CARS signals are generated through the interaction of laser fields under the condition of phase matching. The induced polarization in the sample depends on both resonant and nonresonant components of the third-order nonlinear susceptibility. The resonant term is associated with specific molecular vibrations, while the nonresonant term contributes to the background [3, 4, 6, 9]. CARS intensity is proportional to the square of the nonlinear polarization, providing enhanced sensitivity. However, the

R. Sharma, *Nonlinear Optical Materials for All-Optical Switching Applications*, Progress in Optical Science and Photonics 33,
https://doi.org/10.1007/978-981-96-5921-0_10

nonresonant background can diminish contrast and spectral selectivity. Advances in excitation schemes, such as polarization-sensitive detection and time-resolved techniques, have been developed to suppress this background, thereby improving the imaging quality.

One of the major problems in chemical imaging is associated with the photobleaching of the carriers in the excited state owing to the higher power of the incident beam. Also, the fluorophore labels are created to image the molecule, which can induce the perturbation of the cell functions. The resolution of the infrared microscopy is limited at longer wavelengths, and also, increased IR absorption by water content reduces the quality of the cell images. In the case of spontaneous Raman scattering, the shorter wavelengths are used to obtain the images. Nevertheless, the high incident beam power is a primary requirement due to the reduced cross-sectional area in the Raman scattering. Not only does the high incident power requirement, but also the long exposure time in the Raman scattering limit the resolution of images of the living systems. The above-listed problems can be overcome by using CARS, which is a four-wave mixing technique, as depicted in Fig. 10.1. The interaction of pump (ω_p), Stokes (ω_s) and probe beams (ω'_p) produces an antistokes beam with frequency, $\omega_{as} = \omega_p + \omega_s + \omega'_p$. The pump and probe beams are obtained from the single laser source. In some of the experiments, the continuous laser beam is also used as a probe. The CARS is a third-order nonlinear optical process which can be written as

$$P^3 = \chi^{(3)} E_p E_s \tag{10.1}$$

where $\chi^{(3)}$, E_p and E_s are, respectively, the third-order nonlinear susceptibility, pump field and Stokes field.

The general expression of $\chi^{(3)}$ is

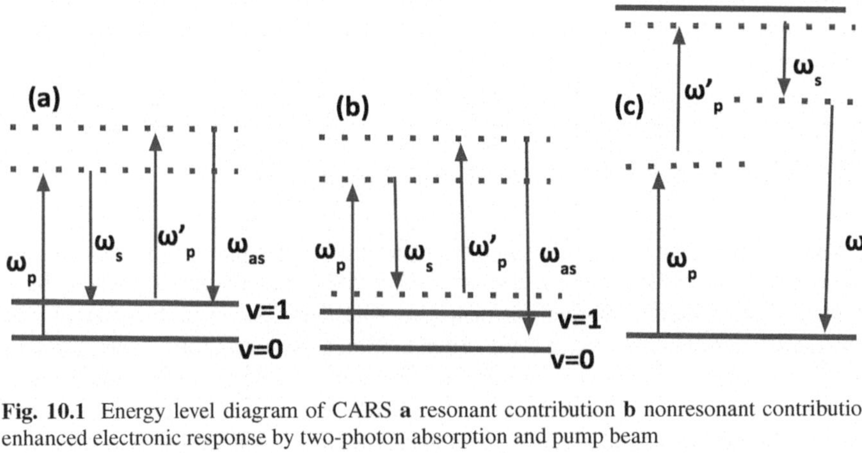

Fig. 10.1 Energy level diagram of CARS **a** resonant contribution **b** nonresonant contribution **c** enhanced electronic response by two-photon absorption and pump beam

$$\chi^{(3)} = \frac{A_g}{\Omega - (\omega_p - \omega_s) - \iota\Gamma_R} + \chi_{nr}^{(3)} + \frac{A_t}{\omega_1 - 2\omega_p - \iota\Gamma_t} \tag{10.2}$$

where Ω, A_R, A_t are vibrational frequency, Raman scattering cross section, and two-photon absorption cross section, respectively. The first term is related to resonant contribution due to vibration frequency, which is also included in Fig. 10.1a. The second term is related to nonresonant contribution and is represented by Fig. 10.1b. The nonresonant contribution is increased due to the two-photon absorption resonance, which is indicated by the third term of Eq. 10.2 and situation is highlighted by Fig. 10.1b.

10.3 Experimental Design

10.3.1 Laser Sources and Pulse Parameters

CARS microscopy typically employs picosecond or femtosecond pulsed lasers. The choice of pulse duration affects spectral resolution and signal strength. Narrowband picosecond pulses offer high spectral resolution and reduced background, making them ideal for imaging specific vibrational modes. Alternatively, broadband femtosecond pulses enable multiplex imaging, capturing entire vibrational spectra in a single acquisition. The synchronization of pump and Stokes pulses is an essential factor for optimal signal generation. Recent developments in laser technology, including jitter reduction and pulse amplification, have significantly enhanced the efficiency and sensitivity of CARS microscopy.

10.3.2 Beam Geometries and Detection

CARS microscopy employs tightly focused laser beams to achieve high spatial resolution. Two primary detection schemes are used: forward detection (F-CARS) and backward detection (E-CARS) [1]. F-CARS is suitable for large or bulk samples, while E-CARS excels in imaging small particles and interfaces due to its sensitivity to localized signals. Polarization-sensitive detection further enhances contrast by exploiting differences in polarization properties between resonant and nonresonant signals. This approach allows selective imaging of specific molecular vibrations, providing high spectral selectivity and reduced background interference.

The forward and backward detection arrangements of the CARS microscopy are highlighted in Fig. 10.2, where spatially overlapped pump and Stokes laser beams are directed to the objective lens through a dichroic mirror. The objective lens focuses the laser beams onto the sample. Then the CARS signal is separated from the excitation beam by using a band-pass filter. Finally, the forward CARS signal (F-CARS) is collected by the detector. On the other hand, the excitation beam is reflected, and

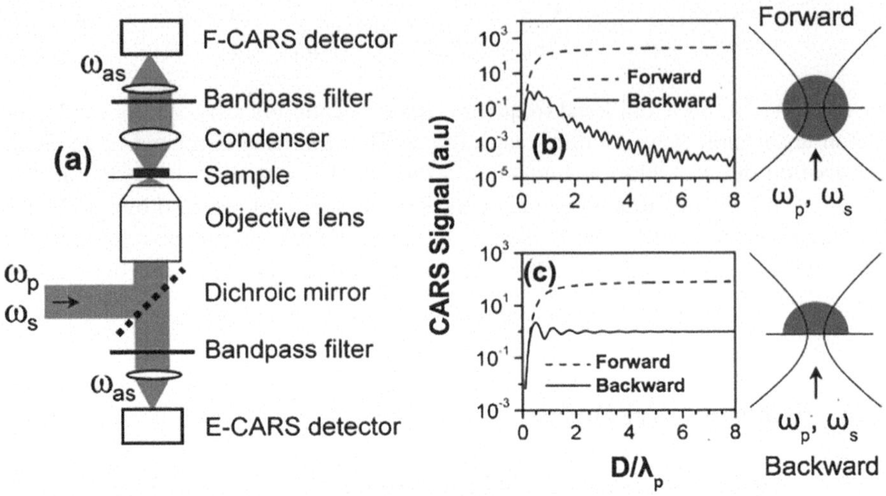

Fig. 10.2 **a** CARS microscopic experimental configuration with forward and backward detection arrangement. The CARS signal when the hemispherical sample **b** at focus and **c** z>0 region. Adapted with permission from [1] Copyright 2004 American Chemical Society

the remaining signal is transmitted for EPI detection. When the pump and Stokes beam have parallel polarization, the calculated F-CARS and E-CARS signals can be represented by Fig. 10.2b. One can see that the F-CARS and E-CARS signals have different profiles. For example, when $D = 8\lambda_p$, the F-CARS signal is about 10^5 times larger than the backward CARS signal. Nevertheless, when the interface is perpendicular to the optical axis, the calculated signal is displayed by Fig. 10.2c. In this case, the separation between the F-CARS and E-CARS signals is reduced.

10.3.3 Multiplex and Time-Resolved Techniques

Multiplex CARS (M-CARS) combines narrow-band and broadband laser sources to simultaneously capture vibrational spectra across a wide range of frequencies. This technique enables rapid spectral characterization of microscopic samples, distinguishing multiple chemical species within a single scan. Time-resolved CARS introduces temporal delays between pump, Stokes, and probe pulses, effectively separating resonant signals from instantaneous nonresonant contributions. This method enhances contrast and allows the study of molecular dynamics with ultrafast temporal resolution.

10.4 Applications in Chemical and Biological Imaging

CARS microscopy has demonstrated sensitivity to lipid molecules due to their abundant C–H bonds. Lipid-rich structures, such as membranes and droplets, can be visualized with high contrast and resolution. For instance, CARS imaging of phospholipid bilayers has revealed the organization of lipids in cellular membranes and the dynamic processes of lipid metabolism. On the other hand, CARS microscopy has been used to monitor lipid droplet formation during adipocyte differentiation. The laser frequencies can be tuned to resonate with the vibration of the C–H bond to observe the growth and distribution of triglyceride droplets in real time, which can provide details of lipid storage and metabolism. The vibrational modes of proteins and nucleic acids have unique spectral signatures that can be detected by CARS microscopy. Protein-rich regions, such as cytoplasmic organelles, have been imaged using the amide I vibrational band. Similarly, DNA in chromosomes has been visualized by targeting the phosphate backbone's vibrational modes. Polarization-sensitive CARS (P-CARS) has been particularly effective in imaging proteins, as it suppresses nonresonant background signals. This technique has revealed detailed protein distributions within cells, which can help to study cellular functions and processes. CARS microscopy is well suited for studying dynamic biological phenomena due to its high-speed imaging capabilities. For example, the progression of apoptosis in fibroblast cells has been monitored using CARS. By imaging lipid-rich membrane structures, researchers have observed morphological changes associated with the early and late stages of cell death.

In another application, CARS has been employed to investigate intracellular water dynamics, providing quantitative measurements of diffusion rates and membrane permeability. These studies highlight the potential of CARS microscopy to probe cellular physiology noninvasively. Beyond biological systems, CARS microscopy has been applied to map the chemical composition of complex materials. In studies of emulsions and gels, CARS imaging has revealed the distribution of water, lipids, and other components, offering insights into phase behavior and molecular interactions. In material sciences, CARS has been used to investigate the orientation and ordering of molecules in structured systems, such as lipid bilayers and polymer films. The polarization sensitivity of CARS provides valuable information about molecular alignment and symmetry, contributing to a deeper understanding of material properties.

10.5 Advanced Techniques and Innovations

Several strategies have been developed to address the challenge of nonresonant background in CARS microscopy. E-CARS and counter-propagating CARS (C-CARS) exploit phase-matching conditions to selectively enhance resonant signals.

Polarization-sensitive detection further improves contrast by isolating resonant contributions based on their polarization properties. Time-resolved and phase-controlled excitation methods provide additional tools for background suppression. These techniques leverage temporal and spectral manipulations of laser pulses to enhance signal specificity, enabling high-quality imaging in challenging samples. CARS correlation spectroscopy (CARS-CS) extends the capabilities of CARS microscopy to study fast diffusion dynamics with vibrational selectivity. By analyzing intensity fluctuations in the CARS signal, CARS-CS provides quantitative measurements of diffusion coefficients, particle concentrations, and medium viscosity. This technique has been applied to investigate the transport properties of nanoparticles and biomolecules in complex environments.

10.6 Summary

CARS microscopy can be considered a significant advancement in vibrational imaging by combining high sensitivity, speed, and three-dimensional resolution. The applications of CARS microscopy span a wide range of fields, from cell biology to material sciences, providing unique insights into molecular structures and dynamics. With ongoing advancements in instrumentation and methodology, CARS microscopy is at the forefront of chemical and biological imaging, driving discoveries and innovations across disciplines. The continuous development of CARS microscopy promises to unlock new applications and capabilities. Advances in laser technology, including the development of compact, cost-effective sources, will expand accessibility and enable broader adoption. CARS microscopy can be coupled to fluorescence and electron microscopy, which can produce images of biological and chemical specimens. Emerging applications in medical diagnostics, such as cancer detection and metabolic imaging, highlight the clinical potential of CARS microscopy. By providing label-free, high-resolution imaging of molecular signatures, CARS offers a noninvasive approach to diagnosing diseases and monitoring therapeutic interventions.

References

1. J.-X. Cheng, X.S. Xie, Coherent anti-stokes Raman scattering microscopy: instrumentation, theory, and applications (2004)
2. E.M. Fantuzzi, S. Heuke, H. Rigneault, Wide-field coherent anti-stokes Raman scattering microscopy using random illuminations. Nat. Photon. **17**, 1097–1104 (2023)
3. M. Garcia, D. Thompson, S. Lee, A novel approach to enhance sensitivity in coherent anti-stokes Raman spectroscopy. Anal. Chem. **96**(2), 1234–1242 (2024)
4. S. Kim, J. Lee, H. Park, Real-time monitoring of chemical reactions using coherent anti-stokes Raman scattering spectroscopy. J. Raman Spectrosc. **52**(4), 789–797 (2021)
5. H. Lee, M. Kim, Y.K. Park, Label-free imaging of lipid droplets in live cells using coherent anti-stokes Raman scattering microscopy. Biomed. Opt. Exp. **13**(2), 456–467 (2022)

6. W. Liu, X. Chen, J. Wang, Advancements in coherent anti-stokes Raman scattering microscopy for dermatological applications. Biomed. Opt. Express **13**(8), 4123–4135 (2022)
7. T. Lv, B. Han, M. Yan, Z. Wen, K. Huang, K. Yang, H. Zeng, Ultra-high-speed coherent anti-stokes Raman spectroscopy with a hybrid dual-comb source (2023). arXiv:2304.08005
8. T. Lv, B. Han, M. Yan, Z. Wen, K. Huang, K. Yang, H. Zeng, Ultrahigh-speed coherent anti-stokes Raman spectroscopy with a hybrid dual-comb source. ACS Photon. **10**(8), 2964–2971 (2023)
9. A. Rodriguez, L. Martinez, E. Gonzalez, Coherent anti-stokes Raman scattering microscopy for rapid histopathological imaging. J. Biomed. Opt. **28**(3), 036005 (2023)
10. T. Smith, E. Brown, M. Jones, Development of a compact coherent anti-stokes Raman scattering microscope for in vivo imaging. Opt. Lett **45**(10), 2760–2763 (2020)
11. L. Wang, Y. Liu, H. Zhang, High-speed coherent anti-stokes Raman scattering imaging with a synchronized dual-comb source. Opt. Exp. **31**(5), 7890–7901 (2023)
12. B. Yao, F. Lin, Z. Luo, Q. Chen, D. Lin, Z. Yang, J. Li, Q. Junle, Recent progress in deep learning for improving coherent anti-stokes Raman scattering microscopy. Laser Photon. Rev. **18**(11), 2400562 (2024)
13. W. Zhang, J.-X. Cheng, Advances in coherent anti-stokes Raman scattering microscopy for biological imaging. Appl. Spectrosc. **75**(6), 611–629 (2021)

Chapter 11
Multiphoton Microscopy and Nonlinear Optical Techniques for Bio-Imaging Applications

11.1 Fundamentals of Multiphoton Microscopy

Multiphoton microscopy (MPM) is an advanced imaging technique that enables high-resolution, three-dimensional visualization of living tissues. Unlike traditional fluorescence microscopy, which relies on single-photon excitation, MPM utilizes the simultaneous absorption of two or more photons to excite fluorophores [14, 18, 20, 23]. This nonlinear process confines excitation to the focal plane, minimizing out-of-focus light and reducing photodamage to surrounding tissues. Consequently, MPM is particularly advantageous for imaging thick, scattering specimens such as brain slices, embryos, and entire organs. The most common form of MPM is two-photon excitation microscopy, where two photons of longer wavelengths (typically in the near-infrared range) are absorbed simultaneously to excite a fluorophore. This approach benefits from deeper tissue penetration and reduced scattering, as longer wavelengths are less prone to scattering and absorption by biological materials [10, 13, 16, 17]. Additionally, the localized excitation inherent to multiphoton absorption decreases photobleaching and phototoxicity, preserving the viability of living specimens during prolonged imaging sessions.

In Fig. 11.1a, the difference between ground and excited energy states of the fluorophore is covered with the help of the two photons of equal frequencies, ω_{ex}. Two-photon fluorescence (TPF) signal is obtained when the transition occurs between the excited state and the ground state. The second-harmonic generation process where frequency 2 ω is emitted when a laser beam of frequency ω is incident on the fluorophore (Fig. 11.1b). TPF signal is recorded from a specimen of the kidney of the mouse, which is shown in Fig. 11.1c. Blue, green, and red colors are, respectively, for DAPI, Alexafluor 488 and Alexafluor 568 dyes [12]. The second-harmonic generation signal image from collagen in the chicken skin is indicated by Fig. 11.1d. In addition to the two-photon excitation, MPM encompasses various techniques, including three-photon microscopy, second-harmonic imaging microscopy, and coherent Raman scattering microscopy [3, 15, 19]. Each of these techniques offers unique

R. Sharma, *Nonlinear Optical Materials for All-Optical Switching Applications*,
Progress in Optical Science and Photonics 33,
https://doi.org/10.1007/978-981-96-5921-0_11

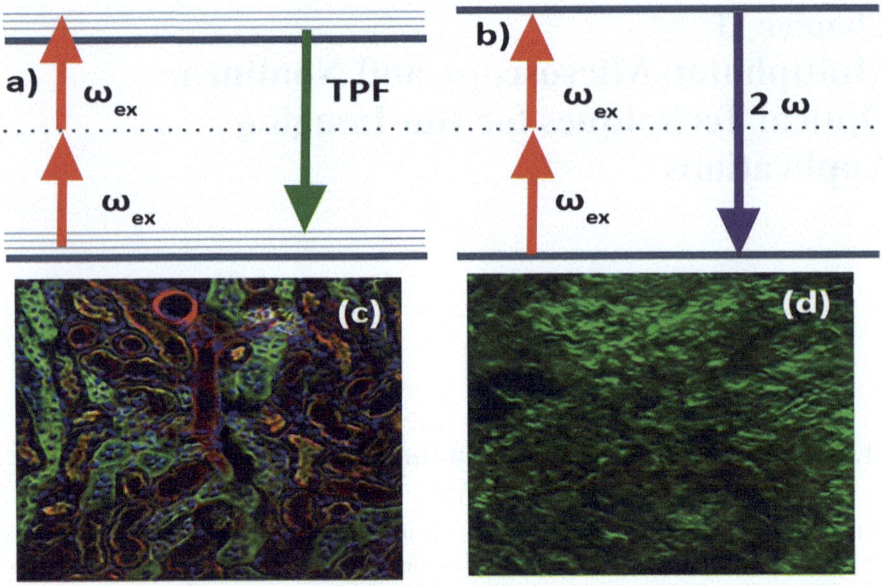

Fig. 11.1 **a** The difference between ground and excited energy states of the fluorophore is covered with the help of the two photons of equal frequencies, ω_{ex}. Two-photon fluorescence (TPF) signal is obtained when the transition occurs between the excited state and the ground state. **b** The second-harmonic generation process, where frequency 2 ω is emitted when a laser beam of frequency ω is incident on the fluorophore. **c** TPF signal from a specimen of kidney of a mouse. Blue, green, and red colors are, respectively, for DAPI, Alexafluor 488, and Alexafluor 568 dyes. **d** The second-harmonic generation signal image from collagen of the chicken skin. Adapted with permission from [12]

advantages for specific applications. For instance, three-photon microscopy provides even greater imaging depths due to the use of longer excitation wavelengths, while second-harmonic imaging enables label-free visualization of non-centrosymmetric structures like collagen fibers.

11.2 Technological Advancements in Multiphoton Microscopy

Recent years have witnessed significant technological advancements in MPM, enhancing its capabilities and expanding its applications in biomedical research. One notable development is the integration of ultrafast fiber lasers as excitation sources. These lasers offer tunable wavelengths, high peak powers, and adjustable repetition rates, providing flexibility for optimizing imaging parameters. The research in the field of MPM has significantly enhanced after the invention of the ultrashort fiber lasers in comparison to the Ti-sapphire based chirped pulse amplifiers. The optical alignments are pretty easy with the fiber lasers. The femtosecond pulsed fiber laser system has been illustrated by Fig. 11.2, which can be used for imaging applications. The front view photograph of the laser system is shown in Fig. 11.2a, which also shows the compactness and portability of the fiber laser system. MÂ^2 beam propagation with <1.1 for both X/Y-axis and symmetrical shape is indicated by Fig. 11.2b while the Typical spectrum is illustrated by Fig. 11.2c. The interferometric autocorrelation of the laser is highlighted in Fig. 11.2d. For example, a study demonstrated that a wavelength-tunable femtosecond fiber laser operating within the 760–800 nm range significantly improved image brightness and penetration depth in two-photon excited fluorescence microscopy. Another advancement is the combination of MPM with light-sheet illumination, resulting in techniques like two-photon excitation light-sheet fluorescence microscopy (TPE-LSFM). This approach enables rapid, high-contrast imaging of large biological specimens with minimal photodamage. Notably, TPE-LSFM has been applied to light-sensitive behavioral studies requiring imaging of extensive regions, such as entire organisms, facilitating comprehensive analyses of complex biological processes. Adaptive optics have also been incorporated into MPM systems to correct for optical aberrations caused by tissue heterogeneity. By dynamically adjusting the imaging system to compensate for distortions, adaptive optics enhance image resolution and signal intensity, particularly when imaging deep within scattering tissues. This improvement is crucial for applications like neuroscience, where high-resolution imaging of neuronal structures in intact brains is essential.

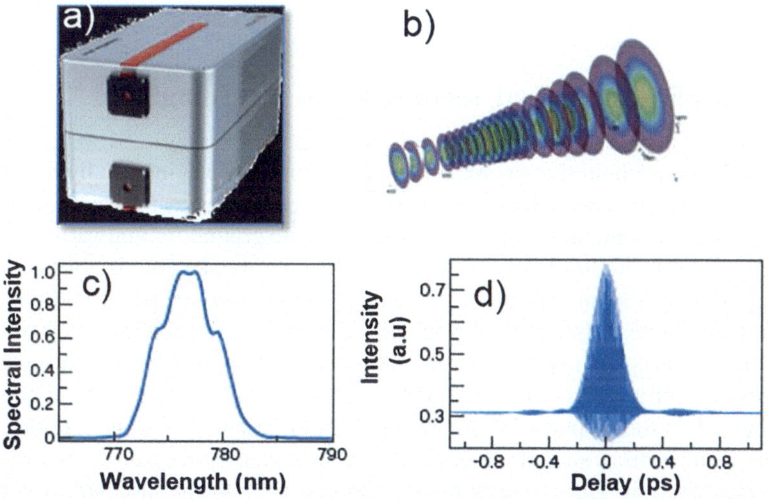

Fig. 11.2 Femtosecond pulsed laser system for imaging applications **a** front view photograph of the laser system **b** MÂ^2 beam propagation with <1.1 for both X/Y-axis and symmetrical shape **d** Interferometric autocorrelation of laser **c** Typical spectrum (Reference: Toptica)

11.3 Applications of Multiphoton Microscopy in Biomedical Research

MPM has become an indispensable tool in various fields of biomedical research due to its ability to provide detailed, three-dimensional images of living tissues with minimal invasiveness. In neuroscience, MPM facilitates the monitoring of neuronal activity in awake, behaving animals over extended periods. By combining MPM with calcium indicators, researchers can visualize and analyze neuronal network dynamics, contributing to a deeper understanding of brain function and disorders. In developmental biology, MPM enables high-content investigation of embryonic development. Recent advances, including optimized laser scanning and label-free strategies, have opened new opportunities for embryo imaging. MPM allows researchers to observe cellular processes in real time within developing embryos, providing insights into morphogenesis and differentiation. Furthermore, MPM has been applied in oncology for the diagnosis and monitoring of tumor progression. By combining MPM with nanomaterials, such as gold nanoparticles and carbon dots, researchers have enhanced imaging contrast and specificity. This combination aids in detecting tumor occurrence, assessing metastasis, and evaluating therapeutic responses, thereby contributing to improved cancer management strategies. The multiphoton microscopy has revolutionized biological imaging by providing high-resolution, three-dimensional visualization of living tissues with minimal photodamage. Ongoing technological advancements continue to enhance its capabilities, broadening its applications across various domains of biomedical research.

11.4 Nonlinear Optical Techniques in Bio-Imaging

Nonlinear optics, a field that explores the interaction between intense light and matter, has become an invaluable tool in bio-imaging. Biological tissues, with their inherent nonlinear optical properties, provide a fertile ground for innovative imaging techniques. Unlike traditional microscopy, nonlinear optical methods offer label-free imaging, high spatial and temporal resolution, and deep-tissue penetration, making them indispensable in biomedical research. This chapter delves into the principles of nonlinear optics, its material basis, and its myriad applications in bio-imaging, from probing cellular structures to diagnosing diseases. Nonlinear optical phenomena occur when the response of a material to light becomes nonlinear under high-intensity illumination, such as laser beams. The induced polarization in a material can be expressed as: Here, represents the nth-order nonlinear susceptibility, which governs second-order effects like harmonic generation and third-order effects such as four-wave mixing. Two categories of nonlinear processes dominate bio-imaging: Second-Order Nonlinear Effects, which include second-harmonic generation (SHG) and sum-frequency generation, occur only in non-centrosymmetric materials. Third-Order Nonlinear Effects include third-harmonic generation (THG) and coherent anti-Stokes Raman scattering (CARS), which are not constrained by material symmetry. Second-Order Nonlinear Materials Second-order effects require materials with non-centrosymmetric structures, such as birefringent crystals (e.g., lithium niobate and potassium titanyl phosphate). Recent advancements have introduced nanostructured and organic materials such as Zinc oxide (ZnO), gallium nitride (GaN) nanowires and nanocrystals, and two-dimensional materials like molybdenum disulfide. Peptide-based nanostructures and self-assembled films that exhibit robust SHG signals can be categorized as organic compounds. Third-Order Nonlinear Materials Third-order effects rely on materials with large susceptibilities, where Silicon nanowires and cadmium sulfide microwires can be used to obtain signals from THG for enhanced THG.

11.5 Super-Resolution Microscopic Imaging

Stimulated Emission Depletion (STED) microscopy employs a femtosecond pulsed laser system as the excitation beam, while the depletion laser beam is obtained by stretching the femtosecond pulsed laser beam to the picosecond regime with the help of optical fibers or gratings. The theoretical idea of the STED microscopy was basically introduced in 1994 by Stefan Hell and Wichmann [8] and was experimentally demonstrated in 1999 [11]. The STED microscopy has made many advancements in the past few years and achievable resolution has been increased which can easily be seen by comparing the confocal and the STED microscopic images of Neurofilament in human neuroblastoma illustrated in Panel (a) and (b) of Fig. 11.3, respectively [6]. In the case of organic dyes, the optical resolution was found to be 150 and 106 nm by using the diode laser and Ti:Sapphire pulsed laser, respectively, which was quite high in comparison to the existing diffraction limit of 200 nm [11, 21].

STED microscopy was implemented for multicolor imaging where super-continuum pulsed laser sources were used to obtain excitation and depletion mechanisms while performing the enhanced resolution imaging of the microtubular network of mammalian cells with a spatial resolution of 30 to 50 nm in the focal plane [22]. STED microscopy was demonstrated with two-photon excitation (TPE) where TPE was carried out using a short-pulse laser source and a CW laser source to obtain a resolution enhancement of about five-fold beyond the diffraction barrier in biological specimens. This study was applied for imaging of brain slices [17]. STED and reversible saturable optical fluorescence transition (RESOLFT) nanoscopic techniques have taken the limits of optical image resolution down to the nanometer scale and opened new paths of biomedical research. Since an intense depletion beam is required to obtain the off states for STED imaging so some of its light can be absorbed and thus can cause photobleaching as well as damage to live cells. In 2011, Gated STED was introduced as a solution to this problem, which provided sharper images at lower power of the STED beam.

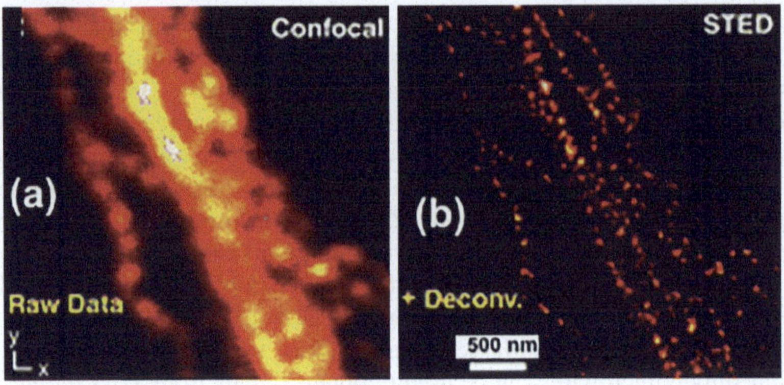

Fig. 11.3 Neurofilament images in human neuroblastoma recorded by using **a** the confocal microscopy and **b** the STED microscopy displaying a focal plane resolution of 20 to 30 nm. Here, the STED image is obtained by applying nonlinear deconvolution [6]

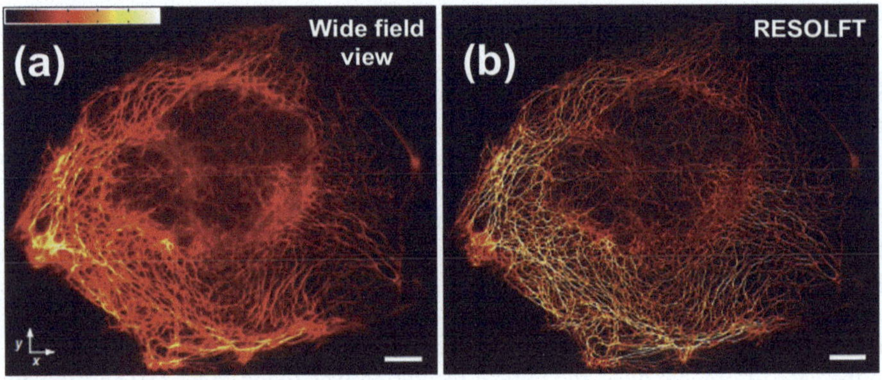

Fig. 11.4 Live-cell images of PtK2 cell expressing keratin 19rsEGFP(N205S) by using **a** wide field view and **b** RESOLFT microscopy [2]

The next advancement in super-resolution imaging is RESOLFT microscopy, which uses a less intense depletion beam for scanning or gated detection and causes less damage to delicate biological samples. RESOLFT microscopy was proposed in 1995 [7]and was implemented in 2007 [1]. These microscopic techniques produce high-resolution images without any typical requirement of image processing or reconstruction, which makes them suitable for live cell imaging in the field of cell biology, microbiology, and neurobiology. In one of the examples, the live cell images of PtK2 cell expressing keratin are depicted in Fig. 11.4, where the RESOLFT microscopic view has shown enhanced optical resolution compared to the wide field view. Also, the light-driven modulation of the fluorescence from ZnSe: Mn nanocrystal were carried out through the ESA mechanism [9]. This leads to a new type of far-field fluorescence microscopy of quantum dots with diffraction-limited resolution. In the literature, there are reports dealing with pulsed laser sources to obtain the excitation and depletion mechanisms. Willig et al. [4] proposed the idea to demonstrate the STED microscopy with a continuous wave laser beam, which can be used for three-dimensional structure imaging of diamond crystal [18]. In order to reduce the complexity as well as cost, the researchers have reported building STED and RESOLFT microscopes with the CW laser sources [20]. Moreover, the researchers had also reported the use of water immersion objective and a spatial light modulator for Z-STED measurements of biological specimens at greater depth [5].

11.6 Summary

The present chapter highlighted an introduction to the two-photon fluorescence (TPF) and its applications to record images of the biological specimen. The potential of nonlinear optics in bio-imaging is vast, with promising directions in which Super-Resolution Imaging nanometer-scale resolution, is achieved by using nonlinear techniques.

References

1. S. Bretschneider, C. Eggeling, S.W. Hell, Breaking the diffraction barrier in fluorescence microscopy by optical shelving. Phys. Rev. Lett. **98**(21), 218103 (2007)
2. A. Chmyrov, J. Keller, T. Grotjohann, M. Ratz, E. d'Este, S. Jakobs, C. Eggeling, S.W. Hell, Nanoscopy with more than 100,000'doughnuts. Nat. Methods **10**(8), 737–740 (2013)
3. M.L. Garcia, D.W. Thompson, S.J. Lee, A novel approach to enhance resolution in nonlinear optical microscopy. Appl. Phys. Lett. **120**(2), 023701 (2022)
4. K.Y. Han, K.I. Willig, E. Rittweger, F. Jelezko, C. Eggeling, S.W. Hell, Three-dimensional stimulated emission depletion microscopy of nitrogen-vacancy centers in diamond using continuous-wave light. Nano Lett. **9**(9), 3323–3329 (2009)
5. J. Heine, C.A. Wurm, J. Keller-Findeisen, A. Schönle, B. Harke, M. Reuss, F.R. Winter, G. Donnert, Three dimensional live-cell sted microscopy at increased depth using a water immersion objective. Rev. Sci. Instrum. **89**(5), 053701 (2018)
6. S.W. Hell, Far-field optical nanoscopy. Science **316**(5828), 1153–1158 (2007)
7. S.W. Hell, M. Kroug, Ground-state-depletion fluorscence microscopy: a concept for breaking the diffraction resolution limit. Appl. Phys. B **60**(5), 495–497 (1995)
8. S.W. Hell, J. Wichmann, Breaking the diffraction resolution limit by stimulated emission: stimulated-emission-depletion fluorescence microscopy. Opt. Lett. **19**(11), 780–782 (1994)
9. S.E. Irvine, T. Staudt, E. Rittweger, J. Engelhardt, S.W. Hell, Direct light-driven modulation of luminescence from mn-doped znse quantum dots. Angewandte Chemie Int. Edn. **47**(14), 2685–2688 (2008)
10. M. Kim, Y.K. Park, Label-free multimodal imaging with simultaneous two-photon and three-photon microscopy. Biomed. Opt. Exp. **15**(1), 114–126 (2024)
11. T.A. Klar, S.W. Hell, Subdiffraction resolution in far-field fluorescence microscopy. Opt. Lett. **24**(14), 954–956 (1999)
12. A.M. Larson, Multiphoton microscopy. Nat. Photon. **5**(1), 1 (2011)
13. H. Lee, M. Kim, Y.K. Park, Development of high-speed multiphoton microscopy for real-time in vivo imaging. Opt. Exp. **29**(10), 15012–15025 (2021)

14. W. Li, P. Wang, C. Xu, Optical sectioning methods in three-dimensional bioimaging. Light: Sci. Appl. **13**(1), 1677 (2024)
15. A. Rodriguez, L. Martinez, E. Gonzalez, Coherent anti-stokes Raman scattering microscopy for rapid histopathological imaging. J. Biomed. Opt. **28**(3), 036005 (2023)
16. J.D. Smith, E.R. Brown, M.T. Jones, Advances in multiphoton microscopy techniques for biomedical imaging. J. Biomed. Opt. **24**(5), 051005 (2019)
17. M. Tian, J. Song, Z. Zheng, X. Xiaoxue, X. Zhong, F. Wang, Multiphoton microscopy in the biomedical sciences xxv. Proc. SPIE **12345**, 1234501 (2023)
18. B.J. Tromberg, Nonlinear optical imaging modalities in biomedicine. Gen. Eng. Biotechnol. News **44**(2), 39–41 (2024)
19. L. Wang, Y. Liu, H. Zhang, Label-free imaging of lipid droplets in live cells using coherent anti-stokes Raman scattering microscopy. Biomed. Opt. Exp. **11**(2), 456–467 (2020)
20. Y. Wang, B. Liu, L. Ding, C. Chen, X. Shan, D. Wang, M. Tian, J. Song, Z. Zheng, X. Xu et al., Multiphoton super-linear image scanning microscopy using upconversion nanoparticles (2024). arXiv:2403.13436
21. V. Westphal, C.M. Blanca, M. Dyba, L. Kastrup, S.W. Hell, Laser-diode-stimulated emission depletion microscopy. Appl. Phys. Lett. **82**(18), 3125–3127 (2003)
22. D. Wildanger, E. Rittweger, L. Kastrup, S.W. Hell, Sted microscopy with a supercontinuum laser source. Opt. Exp. **16**(13), 9614–9621 (2008)
23. W. Zhang, J.-X. Cheng, Recent advances in nonlinear optics for bio-imaging applications. Opto-Electron. Adv. **3**(2), 200003 (2020)

The manufacturer's authorised representative in the EU is Springer Nature Customer Service Centre GmbH, Europaplatz 3, 69115 Heidelberg, Germany. If you have any concerns regarding our products, please contact ProductSafety@springernature.com

Printed and bound by CPI Group (UK) Ltd, Croydon, CR0 4YY
01/10/2026
02225944-0008